Freshwater Fishes

of the Carolinas,

Virginia, Maryland,

and Delaware

Fred C. Rohde

Rudolf G. Arndt

David G. Lindquist

James F. Parnell

Photographs by James F. Parnell

Freshwater FISHES of the Carolinas, Virginia, Maryland, and Delaware

The University of North Carolina Press *Chapel Hill & London*

Publication of this book was assisted by a grant from the John Wesley and Anna Hodgin Hanes Foundation.

Manufactured in the United States of America

The paper in this book meets the guidelines for permanence and durability of the Committee on Production Guidelines for Book Longevity of the Council on Library Resources.

Library of Congress
Cataloging-in-Publication Data
Freshwater fishes of the Carolinas, Virginia, Maryland, and Delaware / by Fred C. Rohde . . . [et al.] ; photographs by James F. Parnell.
p. cm.
Includes bibliographical references (p.) and index.
ISBN 0-8078-2130-6 (cloth : alk. paper)
ISBN 0-8078-4579-5 (pbk. : alk. paper)
1. Freshwater fishes—North Carolina—Identification. 2. Freshwater fishes—South Carolina—Identification. 3. Freshwater fishes—Virginia—Identification. 4. Freshwater fishes—Maryland—Identification. 5. Freshwater fishes—Delaware—Identification. 6. Freshwater fishes—North Carolina. 7. Freshwater fishes—South Carolina. 8. Freshwater fishes—Virginia. 9. Freshwater fishes—Maryland. 10. Freshwater fishes—Delaware. I. Rohde, Fred C. II. Parnell, James F.
QL627.F7 1994
597.092′975—dc20 93-32535 CIP

99 98 97 96 95 6 5 4 3 2

Contents

Acknowledgments

There are a number of scientific and popular books and articles that consider the fishes of more than, or part of, the Carolinas, Virginia, Maryland, and Delaware. There is none, however, that focuses on this five-state region. We have relied heavily on some of these books and articles as we prepared this book, but we have not cited all of the authors of these studies and singled out their specific contributions, as we would have done in a more technical publication. We acknowledge our debt to them and list the most important and comprehensive works that we consulted under Selected References at the end of this book.

We are also indebted to the following people for their help in the preparation of this book. Robert E. Jenkins of Roanoke College, Salem, Virginia, and Noel M. Burkhead of the United States Fish and Wildlife Service, Gainesville, Florida, provided information on the distribution of the game fishes of Virginia. David A. Etnier of the University of Tennessee, Knoxville, and Robert E. Jenkins reviewed the entire manuscript, and Steve W. Ross of the North Carolina National Estuarine Research Reserve Program, Wilmington, reviewed the introductory chapters; the comments and suggestions of all three have improved the book.

The Graphics Production Department, especially Julie E. Bowen and Lai-Kam Lee, of The Richard Stockton College of New Jersey, Pomona, New Jersey, prepared the final version of all nonphotographic figures. The Research and Professional Development Committee and the administration of The Richard Stockton College of New Jersey kindly made possible released time from teaching responsibilities for Rudolf G. Arndt for work on this book. The University of North Carolina at Wilmington provided assistance with the preparation of the manuscript and with photography.

A. Brett Bragin provided use of certain field equipment and assisted in collecting fishes. Andrea M. Teti typed portions of earlier drafts of the manuscript and assisted in field work. Keith M. Ashley, C. Wade Bales, Richard G. Biggins, Noel M. Burkhead, Clyde E. (Billy) Campbell, Jim Dineen, Steven R. Layman, Stephen P. McIninch, Joe H. Mickey, Jr., Stephen T. Ross, William N. Roston, Christopher W. Sample, J. R. Shute, Peggy W. Shute, and Craig A. Harper provided assistance in the field or donated 35-mm slides or specimens. Thanks go to Hugh Morton at Grandfather Mountain, North Carolina, and to the North Carolina Aquarium at Manteo for permission to photograph fish in their tanks. The South Carolina Wildlife and Marine Resources Department, through John E. Cely, and the North Carolina Wildlife Resources Commission, through Fred A. Harris, kindly issued, over a period of many years, the required scientific research permits that allowed us to collect and study fishes in their states. We thank students and colleagues, too numerous to mention here, who have helped us to collect fishes for the past fifteen years or so.

Finally, special thanks go to our wives, families, and friends, who provided support and encouragement throughout this project.

Freshwater Fishes of the Carolinas, Virginia, Maryland, and Delaware

Introduction

This book, a companion to the three University of North Carolina Press books already available on the amphibians and reptiles, birds, and mammals of the mid-Atlantic states, attempts to familiarize students, amateur naturalists, hobbyists, educators, and fishermen with the fishes that occur in the fresh waters of South Carolina, North Carolina, Virginia, Maryland, and Delaware.

The fish fauna of the southeastern United States is the most diverse of any in North America, and the numbers of species that occur in some portions of western Virginia and western North Carolina are among the greatest for any area in the Southeast. Thirty-nine species occur only in the five-state region considered here. Despite this great diversity—often coupled with an abundance of individuals and an abundance and diversity of aquatic habitats—the average person usually knows only the game fishes. Any fish that is not a bass or sunfish or trout, and often even the young of these, is usually referred to as a "minnow." This book provides basic information on how to identify the fishes of the region, their life histories, where and how to catch them, how to observe and photograph them in the wild, and how to maintain and photograph them in captivity.

The five-state area of coverage is bounded on the east by the Atlantic Ocean and on the west by the Appalachian Mountains. Although the northern and southern boundaries are more politically than naturally defined, they do coincide roughly with breaks in the distribution of some northern species that are near the southern edge of their ranges and of some southern species that are near the northern edge of their ranges.

As might be expected since the region is located in the middle of the Atlantic coast and includes the major physiographic provinces of the eastern United States, most of the fishes found here are also found well outside the region. Consequently, this book will also be of value to persons outside the area on which it concentrates.

Although written with the nonbiologist in mind, the book includes some technical terminology intended to better bring the reader into the world of fishes and of the ichthyologist (one who studies fishes). Generally, technical terms are explained when they are introduced in the text, but a glossary is also provided at the end of the book.

Altogether, 267 species are considered here, of which 262 now occur within the states of coverage. This number includes four species that are not native to North America but now occur in the region, six species that are native to North America but not to the region and are now established here, and five species that have been extirpated from the region or are now extinct. Some 85 of the 262 species are currently of special concern to the federal government and to one or more state governments either because they have exhibited a serious decrease in numbers or distribution or because their status is not known.

Published data on the biology of fishes from populations in the five-state region have been used whenever possible. When necessary, however, such information has been augmented with data collected from outside the region, and in some cases only outside data were available. The authors have included their own observations whenever possible. Please consult the works listed under Selected References at the end of the book for more detail.

The authors hope that this book will increase awareness and understanding of, as well as interest in, the fishes of the Carolinas, Virginia, Maryland, and Delaware and that it will encourage the conservation and the further investigation of these fascinating animals.

Brief History of Ichthyology of the Region

One of the first accounts of the fishes of the mid-Atlantic region was provided by an Englishman, Thomas Harriot of Oxford, in 1588. In that year he published—complete with illustrations by John White—his observations, made on an expedition three years earlier, of the longnose gar and some other fishes from the vicinity of the colony of Roanoke established in North Carolina by Sir Walter Raleigh. Other early anecdotal passages exist as well, but the great early work, not only for the mid-Atlantic region but for American ichthyology in general, is *The Natural History of Carolina, Florida, and the Bahama Islands* by Mark Catesby, published between 1731 and 1743. Although Catesby was not well educated, his figures and descriptions of fishes were more detailed than those of his predecessors.

In 1822, John E. Holbrook, a Charleston, South Carolina, physician, pioneered the study of the fishes of his area. His beautifully illustrated *Ichthyology of South Carolina* was published in 1855 and 1860. He provided the original descriptions of two of the freshwater fishes found in the mid-Atlantic region.

The western part of North Carolina and Virginia received attention from Edward D. Cope, an outstanding early ichthyologist. Cope collected in several of the mountain drainages in Virginia in 1867 and in the foothills and piedmont of North Carolina in the autumn of 1869. He published his findings from these collecting expeditions, respectively, in 1868 and 1870. He was the pioneer ichthyologist of North Carolina and described a total of forty-seven of the fishes currently found in the region.

Perhaps the greatest legacy of Louis Agassiz, another early ichthyologist, aside from his own stalwart work on fishes, was the encouragement he gave to one of his youngest students, David Starr Jordan. Jordan and others working with him investigated some streams of the upper piedmont and mountain regions of North Carolina and Virginia during the summer of 1888, and they reported on the fishes of the headwaters of the Pamlico, Neuse, Cape Fear, Pee Dee, Santee, and French Broad rivers. Fishes of the lower piedmont and coastal plain were much more sparingly studied due to poor roads, swamps, and mosquitoes.

Jordan described thirty of the fish species found in the five-state region. His monumental contribution, *The Fishes of North and Middle America*, co-authored with Barton W. Evermann, was published from 1896 to 1900. This treatise did much to compile and synthesize knowledge of fishes in its area of coverage, so much so that Jordan told his last student, Carl L. Hubbs, that it would be wise for him to work on the fishes of another continent since North American fishes were already known! Indeed, freshwater ichthyology in eastern North America fell dormant for many years after the publication of this masterful work.

Clement S. and Herbert H. Brimley (generally known as the Brimley brothers), who later founded what was to become the North Carolina State Museum of Natural Sciences, collected fishes in the area around Raleigh and elsewhere in North Carolina during the late 1800s. These collections were later used by Barton W. Evermann and

Ulysses O. Cox to describe the fishes of the Neuse River in 1895 and by Barton A. Bean to describe the fishes of the Cane River and Bollings Creek in 1903.

During the 1890s, Hugh M. Smith, with B. A. Bean, published on the fishes of the lower Potomac River, Maryland, and the District of Columbia. In 1907, Smith published *The Fishes of North Carolina*, the first comprehensive work on the marine and freshwater fishes of that state. He summarized the existing studies and added much original information. Much of his work was based on fishes deposited in the North Carolina State Museum by C. S. Brimley. In 1929 Reginald V. Truitt, with B. A. Bean and H. W. Fowler, published on the fishes of Maryland.

During the middle of the twentieth century, Carl L. Hubbs, with associates, described and revised many fish groups of the mid-Atlantic region, including the fishes endemic to Lake Waccamaw, North Carolina. Hubbs described twelve of the species found in the region. Edward C. Raney and his students published a number of papers on the fishes of the region, and they described five new species found here.

Henry W. Fowler, the first curator of fishes at the Academy of Natural Sciences in Philadelphia, made frequent collecting trips to Delaware and Maryland in his early days at the beginning of the twentieth century. His technique later was to contact both sport and commercial fishermen to solicit unusual specimens for his collection. As he acquired material for description, Fowler would study and rapidly put into print his findings, often in the proceedings of the academy. Fowler's most notable relevant contribution was his monograph entitled *A Study of the Fishes of the Southern Piedmont and Coastal Plain*, published by the Academy of Natural Sciences in 1945, which included the fishes of all the states under consideration here except Delaware. Most of the book is devoted to freshwater fishes, many examined by Fowler at the Charleston Museum and the North Carolina State Museum in Raleigh. Unfortunately, when describing new species, Fowler rarely examined relevant specimens at other institutions. This lapse, plus the rapidity with which he published his findings, resulted in many of his new species later being recognized as invalid, although five of the freshwater species he described for the region still stand. Despite Fowler's shortcomings in this regard, the scope and the breadth of his ichthyological achievements have seldom been equaled.

Ichthyology has now become a popular field of study, and courses on fishes are taught in many colleges and universities. Some of the larger universities and colleges also have graduate programs that provide advanced training in fish ecology, behavior, physiology, and natural history. Many graduates of such programs work in a field of applied ichthyology, fisheries biology, and are responsible for the study and management of the extensive recreational and commercial fisheries of the mid-Atlantic region. Ichthyologists and fishery biologists are employed by various private groups and by government agencies or work as professors who teach, conduct research, and provide community and regional services through their colleges and universities. A recent estimate indicates that at least 216 professional fishery biologists or ichthyologists are active in the region today. Using time-tested tools and methods, as well as more modern ones such as genetic engineering and molecular biology, many ichthyologists today follow in the footsteps of Cope, Jordan, Hubbs, Fowler, Raney, and others as they continue to study the description, classification, and natural history of the fishes of the region.

Characteristics of Fishes

Almost everyone can identify a fish by sight, but the group is so old, contains so many species, and exhibits such a huge range of adaptations that it is difficult to define the word "fish." Actually, "fish" refers to several classes of vertebrate (with a backbone) animals that are as different from one another as, for example, amphibians are from reptiles and reptiles are from birds, and a definition for all of the great diversity of animal life included in the word "fish" is not easy. Fish classification, even at its most scientific, is subject to disagreement among professionals. The most favored current system of classification divides the living fishes into four classes: hagfishes (Myxini); lampreys (Cephalaspidomorphi); sharks, skates, rays, and ratfishes (Chondrichthyes); and bony fishes (Osteichthyes). Some biologists prefer to separate the first two groups from the fishes and call them fishlike vertebrates instead.

Generally, we can define a fish as (1) a vertebrate animal that spends all or most of its life in water, breathes by means of hidden gills located in the area of the pharynx, and moves by fins. If we restrict our definition to those freshwater fishes that are found in the region we are examining, we can add that (2) most, but not all, have an internal supportive system comprised of bone; an outside covering of thin, light, and highly flexible scales; two pairs of paired fins and three unpaired fins; four gill arches per side; and paired nostrils.

Fishes are adapted to move rapidly through water, a medium about 800 times as dense as air. This has resulted in the evolution of a fusiform (streamlined) body that reduces resistance on the swimming fish, at least in the postlarval stages of most species. Other life history stages of fishes, such as the eggs and sometimes the larvae, are designed to sink or to catch on to plants or the substrate or to float and be carried along by moving water, and thus these stages often are not streamlined.

Basic fish structures (see Fig. 1) include a head, body, and fins. Just before the tail fin is a caudal peduncle, which is usually thin and constricted from top to bottom and which allows "snap" from the lateral undulations of the body to be transferred to the tail fin for effective forward thrust.

The fin on the back is called the dorsal fin, and it may be single or multiple. If the latter, in the fishes here considered, the fins may be separated or joined, and thus there may be a first (spinous) and a second (soft) dorsal fin. The tail (caudal) fin, like the dorsal fin, is unpaired, as is also the anal fin, located on the ventral midline, just behind the anus. All fishes in the mid-Atlantic region have only one anal fin. It is usually free but is joined to the tail fin and to the dorsal fin in the American eel and in the lampreys. Unpaired fins in fishes in the region function primarily as stabilizers, but in some species they also serve as "billboards" for advertisements and displays, especially in the males.

The paired pectoral fins are located on the side or breast, just behind the head. The pelvic fins are also paired and may be located in the "pelvic" area at the rear of the fish, in the "chest" area, or in the "neck" area, and are then referred to as, respectively, abdominal, thoracic, or jugular in position.

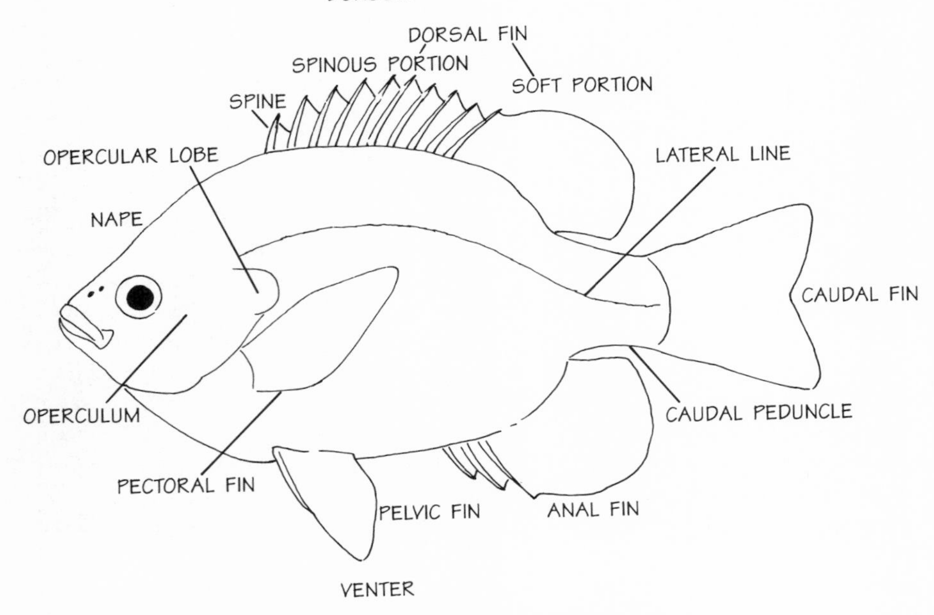

Figure 1. External morphology of a fish, showing commonly occurring structures.

Some fishes—such as trouts, salmons, catfishes, and madtoms—have a small, fleshy adipose fin on the rear of the dorsum and in front of the tail fin. Its function is not known. Unlike other fins, it lacks supporting elements.

The site of attachment of a fin to the body is referred to as its insertion, and the front of the fin where it makes contact with the body is referred to as its origin.

Rays and spines give support to the fins. They can usually be raised or lowered, and they then raise or lower the attached soft fin membranes. Rays are most often soft and flexible, typically branch one or more times, and are segmented (have clearly visible cross-striations). The number of rays present is determined by counting them at the fin base, not at mid- or upper-fin level where they have usually branched.

Spines are unbranched and nonsegmented and are ordinarily hard, rigid, and pointed. Sculpins are an example of the less common fishes that have soft spines. Structures known as false spines are actually hardened, unbranched, and unsegmented rays. Superficially they look and feel like spines, but they are composed of two laterally paired and fused structures. Carp and goldfish have false spines. True spines are single and nonfused structures. All spines provide defense from predators, and many can be locked into an erect position. In catfishes, one or more edges of the spines, which contain venom, are often serrated, and the sharp spine point and serrations can cut readily. The venom can greatly add to the pain that results from being stuck by a spine.

The number of fins, their relative height and length, their color, and their type (with spines or rays, and the number of each), vary from species to species. Thus, details of fin morphology (shape, size, color, etc.) along with body size and other characteristics to be mentioned later are often useful in the identification of species.

Figure 2. Lateral line (holes in the scales) of common shiner, *Luxilus cornutus*.

Fishes in the mid-Atlantic region have a wide range of body shapes. Some are strongly laterally compressed, some weakly so, and some are cylindrical. Some have an elongate body, others a shorter one. Some lack scales, having a body covered with smooth skin, and are referred to as naked. The sturgeons have rows of platelike bony scutes separated by skin, while the armorlike suit of gars is made up of large, thick, inflexible scales. Most fishes are covered with thin, light, flexible, overlapping scales. When such scales have the back or free edge rounded, and the fish feels smooth when it is stroked with the finger tips from tail to head, the scales are referred to as cycloid. When the scales feel sandpapery or rough when stroked, it is because they have tiny, backward-pointing teeth on the rear edge, and they are then referred to as ctenoid scales. All these characteristics are adaptations to particular ways of life and are the product of the evolutionary history of a particular group.

In addition to the usual vertebrate senses such as sight and smell, most fishes have a lateral line system (see Fig. 2), which is unlike any sensory system known in man. Its main receptor is the lateral line, and it is usually located along the midside on both sides of the fish. The system consists of ducts that extend beneath the skin and open to the outside through a pore in the scales, or through the skin if scales are lacking. The ducts are equipped with free nerve endings located at intervals. Branches extend to parts of the head, and the branches on the two sides of the fish are often connected across the back of the head. The system is highly sensitive to water pressure and can best be compared to our auditory system, which is highly sensitive to, and interprets the meaning of, airwaves that impinge on our ears. The intensity and duration of nerve stimulation is interpreted, in concert with other data such as taste, smell, sight, and touch, by the receiver and its biological significance determined by the fish, after which the fish can make an appropriate response. A rival male, an approaching fe-

Figure 3. Plicae on lower lip of brassy jumprock, *Scartomyzon* new species.

male, fleeing prey, a rock falling in water, a crayfish digging a burrow, etc., all create distinctive underwater compression waves, the significance of which is interpreted by a given fish species.

The number, size, and placement of the pores is unique to a species. For example, the lateral line can be straight, curved up or down, run the length of the body (be complete) or not run the length of the body (be incomplete), extend onto the tail, or be absent. Combined with other countable characters (e.g., the number of spines, rays, or scales) and proportional characters (e.g., the diameter of the eye relative to the height or length of the head, or the height of the extended dorsal fin relative to the body depth), lateral line details are tools that help biologists to define and identify particular species.

All fishes in the region under consideration have eyes. Their size, location, and color are important to biologists in helping to identify particular species.

All fishes have a mouth. Some fishes (hagfishes and lampreys) lack jaws, but all the rest have them. The mouth can be at the front and top of the head (superior), at the extreme front (terminal), or slightly or totally under the head (inferior). The lips may be thin or thick and equipped with folds of skin (plicae; see Fig. 3) or individual budlike extensions of skin (papillae). The relative size of the mouth and the number, size, and shape of the teeth help to identify species.

In all the region's fishes except the lampreys, the gills are comprised of a bony, semicircular gill arch, on the rear of which are attached two rows of backward-pointing, highly vascularized, and delicate gill filaments. These function in gas exchange. On the front of the arch in most species is attached a row of bony nubbins called gill rakers, which function as a filter of large and coarse items such as sand and gravel that might be ingested during intense feeding or intensive respiration. Such items that can nick and damage the delicate gill filaments are caught on the rakers, after which the fish "coughs" and ejects them from the mouth. Gill rakers also function to retain large food items that were not swallowed after first having been taken into the mouth.

In planktivores such as the paddlefish and herrings, gill rakers have evolved into numerous, long, and thin structures. These form a sieve that functions like baleen in some whales. The shape and size of the gill arch and the number, shape, and size of gill rakers are important tools in species identification.

Most fishes have teeth. A planktivore generally has none or only a few small ones. A piscivore (fish eater), such as a pike, has numerous large, pointed, and similar-sized canine teeth on both jaws. Fishes such as some drums that feed on large and hard prey are equipped with large, rounded teeth, present not only on the jaw edges but also on all of the surface of both jaws. Such pavementlike dentition and associated heavy jaw bones and large muscles provide the hard surface and pressure necessary to crush mollusks and similar hard prey.

Other fishes, such as catfishes, usually have only small, fine teeth on the jaws but may also have teeth on bones in the roof of the mouth. Such villiform teeth are low, numerous, and sharp and assist in feeding on worms, insect larvae, and other slow, soft prey. Many of the region's minnows and suckers have grinding teeth located on the last gill arch in the throat. These are referred to as pharyngeal teeth.

The adult male of many species of fishes develops outgrowths on various parts of the body during the spawning season. These breeding or nuptial tubercles can occur on the head, body, and/or fins (see Fig. 4). They can be large and pointed like a rose thorn and widely separated, or they can be small and pointed or rounded and so numerous that they occur in a sheet. They may be white or cream-colored and may either appear to give the fish a particular texture and color of skin or be almost unnoticeable. Fishes use them in courtship and mating. These are lost afterward, and their bases are then grown over and concealed by skin. Tubercle "spots" are easily discernible on species with recently lost and large tubercles. The combination of size, number, and arrangement of tubercles is unique to a species. Because the tubercles are present only during the spawning season and are not obvious on all species, many persons do not know about them; therefore, when some people see a fish with tubercles, they are perplexed as to what the structures are or what kind of "new" fish they have found.

Fishes also exhibit a great range of colors. Most have species-specific colors by which they identify members of their own and/or other species. Species-specific colors tend to be most developed on a successful spawning male, for mating purposes. Color also helps fishes to school, hide, find food, and reduce competition by avoiding each other.

The female is usually less colorful than the male, although at breeding time her drab colors may brighten. The young often look like a nonbreeding female, but in a number of species the color of the young differs from that of either adult. Some species are extremely colorful, while others are much less brilliant.

Color also varies with the general health of a fish, whether or not it is in spawning condition, whether or not it has just eaten or just been frightened, etc. The brightest colors are found on a fish that is healthy, in spawning condition, and not frightened, and the dullest occur on a fish that is at the opposite end of this spectrum. Color is also subject to individual variation, where one fish, under the best, or the worst, of

Figure 4. Tubercles on male bluehead chub, *Nocomis leptocephalus*. William N. Roston.

Figure 5. Barbel in corner of mouth of thicklip chub, *Hybopsis labrosa*. J. R. Shute.

conditions, will simply be more or less colorful than another individual. When color is used to help identify fishes, the best specimens to use are large breeding males.

Some fishes have other identifying structures, for example, barbels (see Fig. 5). Barbels range in size from large and readily visible to tiny and hidden, and they vary in shape, color, number, and location among species. Thus they are also of taxonomic importance. Catfishes and madtoms have large barbels, which are fleshy extensions around the mouth that act as "feelers," are equipped with taste buds, and help locate food in turbid waters or at night. Some minnows have small barbels. Additional morphological and behavioral differences are discussed later in the species accounts.

Habitat

A prominent ecologist wrote that if the range of a species is the street on which it lives, its habitat is its house number. For example, in one geographic area (the range) some fishes live in large rivers, others in creeks, some in ponds, etc. (the habitat). These habitats can be looked at more discriminatingly, however, so that, for instance, mountain rivers are different habitats from coastal rivers, fast creeks different from slow, and deep lakes different from shallow.

Furthermore, even these major habitats can be subdivided. For example, a given river can usually be subdivided into riffle habitat, pool habitat, and run habitat (see Fig. 6). The first would usually be inhabited by darters and madtoms, the second by sunfishes and basses, and the last by minnows and suckers. And often each of these habitats can be divided further still. Current in riffles may range from relatively slow to fast, depth from shallow to deep, and substrate particle size from small to large. Darters usually occur in a riffle where the water is faster, while madtoms occur where it is slower, and so on.

Different life history stages of a fish—such as the egg, larva, juvenile, and adult—and the different sexes often require different habitats. For example, eggs of the American eel are deposited in deep and salty ocean waters. The young eel, or elver, migrates to shallow brackish water; the older male then lives in brackish water, while the female moves further upstream into fresh water. The habitat requirements of the life history stages of most of the fishes considered here are, however, less pronounced.

Every habitat consists of abiotic (nonliving) and biotic (living or once living) variables and of the complex and always changing interactions between them. Examples of abiotic variables that help to determine the type of habitat are water depth, flow, color, turbidity, pH, and temperature, current speed, substrate type, water body width, and concentrations of dissolved gases, especially of oxygen and carbon dioxide, within the water. Examples of important biotic variables are the absence or presence of vegetation, plankton, other fishes, invertebrates, birds, and mammals; these variables are important to fishes as food, predators, shelter, egg deposition sites, parasites, etc.

There is always a complex relationship within and between abiotic and biotic variables. Thus rocks in a riffle may be bare or may be populated with algae, with vascular plants, or with both, and to varying degrees. The composition and color of the rocks themselves will help determine water quality characteristics, e.g., whether the water is alkaline, acid, or neutral. The rock-plant interactions, and the length of time for which any set of interactions has prevailed, has an impact on the invertebrates present, and thus on the fishes. The species, numbers, and sizes of fishes present in turn affect the presence and success of plant species within the habitat.

Freshwater habitats can be divided into two major categories: lentic, or those with standing waters, such as lakes, ponds, borrow pits, and swamps, and lotic, or those with moving waters, such as rivers and creeks. The presence or absence of a current is the obvious major difference between them, but related to this is the fact that stand-

Figure 6. Stream habitats (riffle, pool, and run) in the Watauga River, North Carolina. J. R. Shute.

ing waters often have lower dissolved oxygen concentrations and larger seasonal temperature extremes when compared to moving-water habitats. The substrate of lentic habitats is often silt, mud, and decomposing vegetation, and that of lotic habitats often rock, gravel, and sand. Fine and soft substrate material in lotic habitats is usually carried away by the current.

Lakes are large bodies of standing water, usually with a surface area of several dozen to tens of thousands of acres (see Fig. 7). There are only a few natural lakes in the five-state region examined here, among them Mountain Lake in western Virginia, Lake Drummond in southeastern Virginia, and Lake Waccamaw and Phelps Lake in eastern North Carolina. Reservoirs are large lakelike but manmade structures that were created to store drinking water, power electric generators, control floods, and provide opportunities for recreation, including fishing. Ponds may range in size from fractions of an acre to several dozen acres, and can be natural or manmade. Borrow pits (see Fig. 8) are abandoned manmade structures and can be small or large, but they usually have a lower productivity of plant and animal life than ponds due to the sterile nature of their sand and gravel substrate. In the mid-Atlantic region, lakes, ponds, and sloughs are usually shallow, relatively warm, and with a fine sand/mud substrate. Consequently, they usually receive many nutrients and are highly productive, supporting an abundance of submerged and emergent aquatic vegetation, and providing shelter for fishes in the form of "weed" beds, tree trunks, and branches.

The word "stream" refers to any flowing body of water and, as used in this book, includes rivers and creeks. Rivers are usually large (long and wide) streams, although a few are only several feet wide. Creeks and their branches may be only a foot or two wide or up to several tens of feet wide, and they are usually characterized by a current, by a sand, gravel, or rock substrate, and by a relative or complete lack of dead trees or branches in the water. Pools, or deeper areas of streams, usually have reduced water

Figure 7. Lake Waccamaw, North Carolina. David G. Lindquist.

Figure 8. Borrow pit, South Carolina. Rudolf G. Arndt.

movement, a mud or organic-material-rich substrate, and warmer water than is found in runs. Runs are faster-moving areas of moderate depth; in riffles the water moves rapidly and removes organic material from the bottom. Any water body in the mountains or piedmont is likely to have a rock, gravel, or sand substrate, while on the coastal plain the bottom is usually of sand and mud.

The effects of human activity have been superimposed in the last few centuries on what was already a tremendous natural diversity in habitat in the region. Reservoirs, borrow pits, and many ponds are manmade, and many rivers and creeks are today man-modified (dammed, channelized, deepened). The use of fertilizers, herbicides, and pesticides and the entry of these chemicals into water affects fishes and other biota. Animal and plant management programs in waters and on lands adjacent to aquatic systems have a profound impact. These and other human activities have been so pervasive that all habitats have been modified to a greater or lesser degree, and truly natural habitats no longer exist.

Zoogeography

Why is the Cape Fear shiner found in only a small section of the Cape Fear River in North Carolina and the sandhills chub in only the small sandhills area of North Carolina and South Carolina, while the spottail shiner and creek chub are found through much of the eastern United States? Why is the Maryland darter found in only one small creek in northeastern Maryland, while the related greenside darter is found throughout many states? In fact, why are thirty-nine species of fishes (see Table 1) found only in the region considered here and nowhere else in the world? Did they evolve here, did they move here from someplace else, or both? When and how did this occur? Such questions have long been studied by ichthyologists, and the study of fish distributions is part of the field of zoogeography.

Nineteen major river drainages occur in the region (see Fig. 9). Of these, sixteen flow eastward into the Atlantic Ocean and three westward into the Ohio River. One of the latter, the Tennessee River, has six major tributary rivers in western Virginia and North Carolina. The distribution of many of the fishes in the region is restricted to one or several of these nineteen different drainages.

Distribution is influenced by many factors, which include past geologic events such as plate tectonics (the movement of the earth's crust) and glaciation, physiography (form and structure of the earth surface), ecological and physiological tolerances of fishes and, more recently, the effect of humans. Another important variable in distribution is chance: many fishes in the region occur where they do because the site in question is the only place to which they had access, not because it is the only place in which they could survive.

Most of the twenty-six families of freshwater fishes found in the mid-Atlantic states also occur on other continents, especially Europe, because of the similar environments in the regions and because of earlier geological connections. A few major groups such as the families of catfishes, cavefishes, pirate perches, sunfishes, and pygmy sunfishes, and a subgroup of the perch family, the darters, are restricted to North America. They either evolved in North America or survived here after they became extinct elsewhere.

The five-state region under consideration can be separated into five distinct physiographic provinces: coastal plain, piedmont, Blue Ridge, ridge and valley, and Appalachian plateau (see Fig. 10). In the species accounts later in this book the latter three provinces are usually referred to simply as "the mountains" or as "mountainous areas." The distribution of many fishes coincides with one of these climatologically and geologically distinct areas, although many species occur in two or more.

The coastal plain in the region includes almost all of Delaware, the eastern and southern portion of Maryland, the eastern third of Virginia and North Carolina, and the eastern and lower half of South Carolina. The coastal plain is a broad, flat region that ranges from the seashore to an elevation of about 500 feet. This province is characterized by sand and sandy soils usually hundreds of feet deep. The coastal plain is often subdivided into an upper, middle, and lower coastal plain, and this division

TABLE 1. *Freshwater Fishes with a Native Distribution Restricted to the Carolinas, Virginia, Maryland, and Delaware and Their State(s) of Occurrence*

	S.C.	N.C.	Va.	Md.	Del.
Carps and minnows					
Greenfin shiner	x	x	–	–	–
Fieryblack shiner	x	x	–	–	–
Highback chub	x	x	–	–	–
Thicklip chub	x	x	x	–	–
Santee chub	x	x	–	–	–
Crescent shiner	–	x	x	–	–
Pinewoods shiner	–	x	–	–	–
Bull chub	–	x	x	–	–
Whitemouth shiner	x	x	x	–	–
Highfin shiner	x	x	x	–	–
Redlip shiner	x	x	x	–	–
Greenhead shiner	x	x	–	–	–
Cape Fear shiner	–	x	–	–	–
Roughhead shiner	–	–	x	–	–
Sandhills chub	x	x	–	–	–
Suckers					
Roanoke hog sucker	–	x	x	–	–
Bigeye jumprock	–	x	x	–	–
Black jumprock	–	x	x	–	–
V-lip redhorse	x	x	x	–	–
Rustyside sucker	–	x	x	–	–
Bullhead catfishes					
Carolina madtom	–	x	–	–	–
Orangefin madtom	–	x	x	–	–
Broadtail madtom	x	x	–	–	–
Topminnows and killifishes					
Speckled killifish	–	x	x	–	–
Waccamaw killifish	–	x	–	–	–
Silversides					
Waccamaw silverside	–	x	–	–	–
Sunfishes					
Roanoke bass	–	x	x	–	–
Pygmy sunfishes					
Carolina pygmy sunfish	x	x	–	–	–
Bluebarred pygmy sunfish	x	–	–	–	–
Perches					
Carolina darter	x	x	x	–	–
Kanawha darter	–	x	x	–	–
Pinewoods darter	x	x	–	–	–
Waccamaw darter	–	x	–	–	–
Riverweed darter	–	x	x	–	–

TABLE I. *Continued*

	S.C.	N.C.	Va.	Md.	Del.
Maryland darter	–	–	–	x	–
Seagreen darter	x	x	–	–	–
Piedmont darter	x	x	x	–	–
Roanoke logperch	–	–	x	–	–
Roanoke darter	–	x	x	–	–

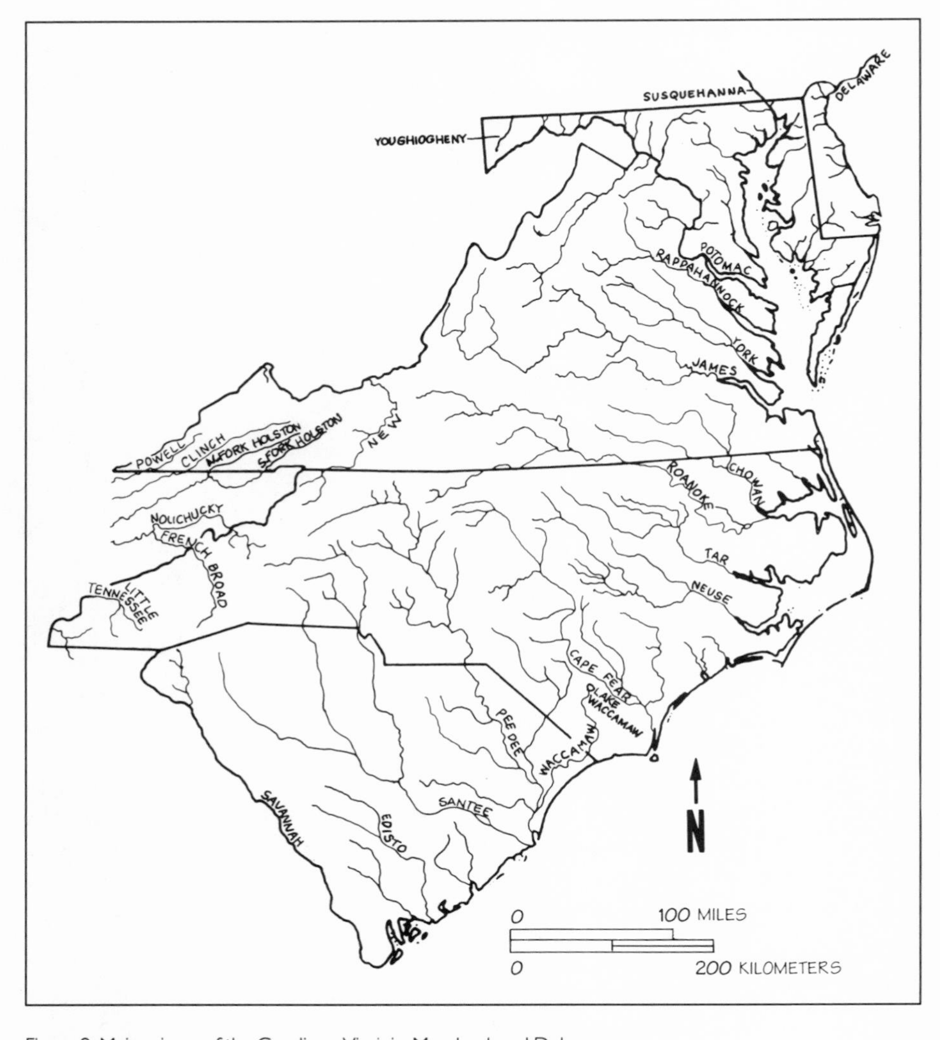

Figure 9. Major rivers of the Carolinas, Virginia, Maryland, and Delaware

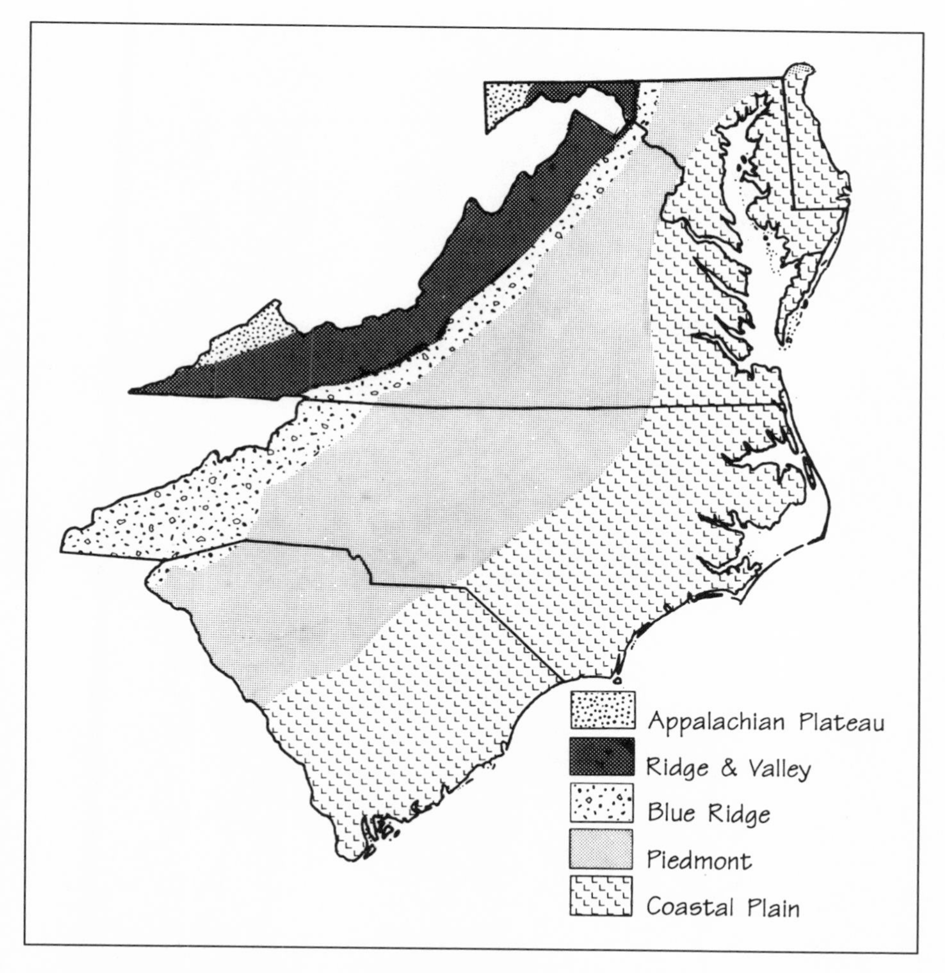

Figure 10. Physiographic provinces of the Carolinas, Virginia, Maryland, and Delaware.

is correlated with elevation. The sandhills in South Carolina and in south central North Carolina occur in the upper coastal plain, which is that portion of the coastal plain located farthest from the ocean. (See Fig. 11.)

The piedmont is located inland from the coastal plain and occupies extreme northern Delaware, the middle quarter of Maryland, central Virginia, the middle third of North Carolina, and most of central and western South Carolina. Elevations in this rolling upland section range from 500 to 2,000 feet. This province is characterized by loose surface soils and/or rocks and is underlain by highly eroded metamorphic rocks. (See Fig. 12.)

The transition zone between the coastal plain and the piedmont is the fall line, which is usually some five to ten miles wide. This is the area where ancient seas deposited sediments that overlapped and covered the older piedmont rocks. In Virginia and Maryland the drop in elevation has fostered the creation of rapids and waterfalls, and thus gave rise to the name "fall" line.

Figure 11. Coastal Plain stream: Nassawango Creek, Maryland. Rudolf G. Arndt.

Figure 12. Piedmont stream: Red Clay Creek, Delaware. Rudolf G. Arndt.

Figure 13. Mountain stream: South Toe River, North Carolina.

The Appalachian Mountains were formed some 280 million years ago when Africa and North America collided, at which time these mountains were taller than the present-day Rocky Mountains. What we see today are the eroded nubbins of these once mighty peaks. This mountain region consists of three conspicuous subdivisions: the Blue Ridge, ridge and valley, and Appalachian plateau provinces. The Blue Ridge province is the easternmost portion of the Appalachian Mountains, with elevations of 2,000 to over 6,000 feet; in the five-state region it occupies a narrow band from western Maryland to northwestern South Carolina. The largest portion and the greatest heights of this province are located in western North Carolina. Within the region, the eastern crest of the Blue Ridge province is generally the divide between waters that flow down the western slope of the Appalachians and then via the Mississippi River toward the Gulf of Mexico and those that flow down the eastern or Atlantic slope and into the Atlantic Ocean. (See Fig. 13.)

On the western side of the Blue Ridge is the ridge and valley province, occupying a part of western Maryland and western Virginia. This is a rugged area of parallel valleys and ridges that runs northeast to southwest at elevations of between 1,000 and 3,000 feet.

The Appalachian plateau province occurs only in extreme western Maryland and extreme southwestern Virginia within the region here examined. It is characterized by plateaus with a moderate to strong relief and an elevation of between 2,000 and 3,000 feet.

The immense age, geologic and climatic stability, and moderate climatic conditions of the Appalachian Mountain area has facilitated the evolution of a large number of fishes, as well as other groups of animals and plants, and accounts for the high species diversity of the mid-Atlantic region. Rivers and streams were formed by water that flowed toward the ocean and by the action of groundwater flows. Some fishes achieved a wide distribution in the region as they moved back and forth through marshy river and creek mouth interconnections along the coast. Others became widely distributed through lateral capture of streams by other streams: as headwater streams of one river system in mountains eroded through rock and entered an adjacent drainage, they sometimes undercut and acquired or "captured" the water of that tributary, and consequently its aquatic biota. For example, four of the rivers that today drain into the Atlantic Ocean—the Susquehanna, Potomac, James, and Roanoke rivers—have cut westward into the Blue Ridge province and consequently intercepted parts of rivers that formerly drained to the west. They thus also captured their biota. The johnny darter and the northern hog sucker, to give two examples, evolved in the Mississippi River basin but through stream capture entered eastward-flowing streams of the James, Roanoke, and other rivers.

Glaciation, erosion, and other physical factors caused the formation and separation of the region's river drainages, often isolating populations of a species. With time, these populations accumulated differences. In general, the longer the isolation, the greater the differences. As a result, a number of species of fishes, as well as several subgenera and genera, have evolved in the region. For example, the seagreen, turquoise, and Swannanoa darters form a distinct subgroup in the darter subgenus *Etheostoma*, and each species occurs in a different drainage. The ancestor of this group is presumed to have evolved in the Mississippi River basin and entered the Santee River drainage through stream capture. The seagreen darter, the most primitive living member of the group, evolved in the Santee River drainage. A population of it reached the Savannah River drainage and there evolved into the turquoise darter. During the same period, other populations of the seagreen darter moved into the Tennessee River drainage and evolved into the Swannanoa darter. All of these changes have occurred relatively recently, as evidenced by the great similarities of the three species.

An example of even more recent evolution is seen in three fishes that are found only in Lake Waccamaw in North Carolina. An unusual habitat, a lake, in this region was formed about 35,000 to 70,000 years ago on the coastal plain. Isolated populations of three common stream fishes—a killifish, a silverside, and a darter—became established there. Adaptations from their former habitat of moving stream waters to quiet lake waters resulted in the evolution of the Waccamaw killifish, Waccamaw silverside, and Waccamaw darter.

Ancient groups of all organisms, including fishes, are usually distributed widely, especially at the level of the higher taxa such as the order, family, and genus. "Newer" forms, such as species, usually have a much more restricted distribution. Rivers and creeks are highways for fish dispersal since fishes are less able to cross land barriers than the more vagile birds and mammals, and, to a lesser extent, the reptiles and amphibians. Waterfalls are an obvious barrier, but so are less conspicuous factors, such as water acidity, temperature, and dissolved oxygen concentration. Some species are adapted to live in a wider range of environmental conditions than others, and these more generalized species usually are better dispersers and are more widely distributed than the more specialized. A saltwater barrier at the lower end of a river usually prevents freshwater fishes from moving from one river into the mouth of another. However, during periods of lower sea levels, such as during glaciations,

Pl. 1. Least brook lamprey, *Lampetra aepyptera*. p. 58.

Pl. 2. American brook lamprey, *Lampetra appendix*. p. 58.

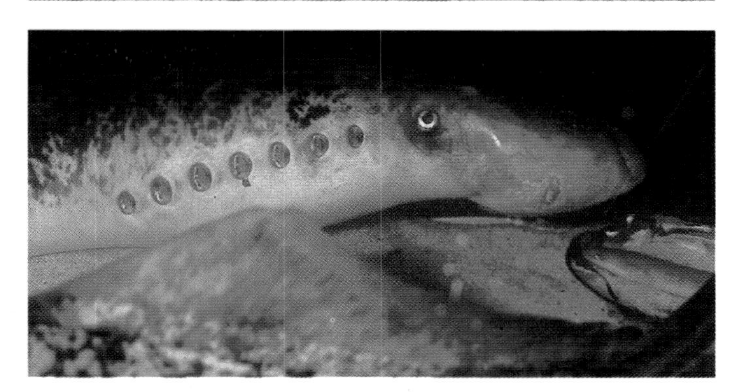

Pl. 3. Sea lamprey, *Petromyzon marinus*. p. 58.

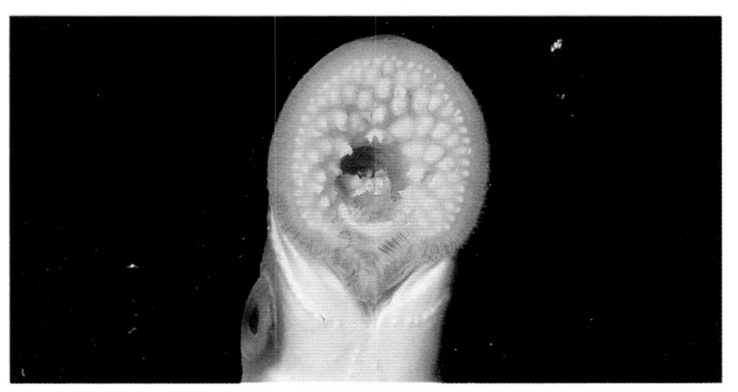

Pl. 4. Oral disc of a sea lamprey, *Petromyzon marinus*. p. 58.

Pl. 5. Atlantic sturgeon, *Acipenser oxyrhynchus*. Noel M. Burkhead. p. 61.

Pl. 6. Paddlefish, *Polyodon spathula*. William N. Roston. p. 63.

Pl. 7. Longnose gar, *Lepisosteus osseus*. p. 65.

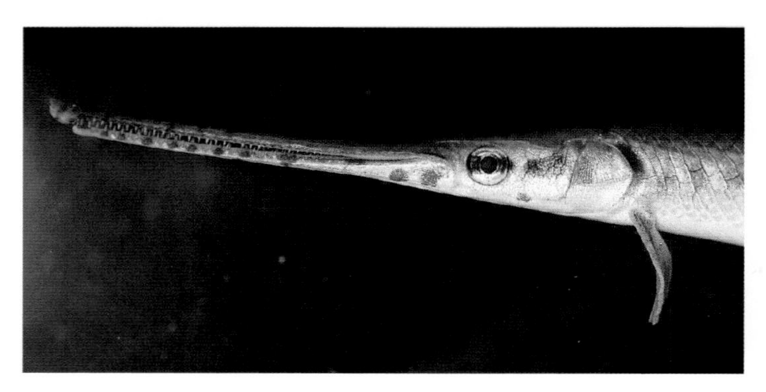

Pl. 8. Bowfin, *Amia calva*. William N. Roston. p. 67.

Pl. 9. Bowfin, *Amia calva*. Juvenile. Fred C. Rohde. p. 67.

Pl. 10. Mooneye, *Hiodon tergisus*. Richard T. Bryant. p. 69.

Pl. 11. American eel, *Anguilla rostrata*. p. 71.

Pl. 12. American shad, *Alosa sapidissima*. p. 73.

Pl. 13. Gizzard shad, *Dorosoma cepedianum*. p. 75.

Pl. 14. Central stoneroller, *Campostoma anomalum*. Male. p. 77.

Pl. 15. Goldfish, *Carassius auratus*. p. 78.

Pl. 16. Rosyside dace, *Clinostomus funduloides*. Male. William N. Roston. p. 79.

Pl. 17. Grass carp, *Ctenopharyngodon idella*. p. 80.

Pl. 18. Satinfin shiner, *Cyprinella analostana*. Male. p. 81.

Pl. 19. Whitetail shiner, *Cyprinella galactura*. Male. p. 81.

Pl. 20. Spotfin chub, *Cyprinella monacha*. Male. Richard T. Bryant. p. 81.

Pl. 21. Whitefin shiner, *Cyprinella nivea*. Male. p. 81.

Pl. 22. Fieryblack shiner, *Cyprinella pyrrhomelas*. Breeding males. p. 81.

Pl. 23. Common carp, *Cyprinus carpio*. William N. Roston. p. 84.

Pl. 24. Silverjaw minnow, *Ericymba buccata*. Richard T. Bryant. p. 85.

Pl. 25. Streamline chub, *Erimystax dissimilis*. Richard T. Bryant. p. 86.

Pl. 26. Blotched chub, *Erimystax insignis*. p. 86.

Pl. 27. Cutlips minnow, *Exoglossum maxillingua*. p. 86.

Pl. 28. Eastern silvery minnow, *Hybognathus regius*. p. 87.

Pl. 29. Highback chub, *Hybopsis hypsinotus*. p. 88.

Pl. 30. Thicklip chub, *Hybopsis labrosa*. Male. William N. Roston. p. 88.

Pl. 31. Rosyface chub, *Hybopsis rubrifrons*. Richard T. Bryant. p. 88.

Pl. 32. Santee chub, *Hybopsis zanema*. Coastal Plain form. p. 88.

Pl. 33. White shiner, *Luxilus albeolus*. Male. p. 89.

Pl. 34. Crescent shiner, *Luxilus cerasinus*. Male. p. 89.

Pl. 35. Warpaint shiner, *Luxilus coccogenis*. Male. William N. Roston. p. 89.

Pl. 36. Common shiner, *Luxilus cornutus*. Male. p. 89.

Pl. 37. Rosefin shiner, *Lythrurus ardens*. Male. p. 91.

Pl. 38. Pinewoods shiner, *Lythrurus matutinus*. Breeding males. p. 91.

Pl. 39. Pearl dace, *Margariscus margarita*. Noel M. Burkhead. p. 91.

Pl. 40. Bluehead chub, *Nocomis leptocephalus*. Male. p. 92.

Pl. 41. River chub, *Nocomis micropogon*. Male. p. 92.

Pl. 42. Golden shiner, *Notemigonus crysoleucas*. p. 94.

Pl. 43. Highfin shiner, *Notropis altipinnis*. p. 96.

Pl. 44. Ironcolor shiner, *Notropis chalybaeus*. p. 95.

Pl. 45. Redlip shiner, *Notropis chiliticus*. Male. p. 95.

Pl. 46. Greenhead shiner, *Notropis chlorocephalus*. Male. William N. Roston. p. 96.

Pl. 47. Dusky shiner, *Notropis cummingsae*. J. R. Shute. p. 95.

Pl. 48. Spottail shiner, *Notropis hudsonius*. p. 96.

Pl. 49. Tennessee shiner, *Notropis leuciodus*. p. 95.

Pl. 50. Yellowfin shiner, *Notropis lutipinnis*. p. 95.

Pl. 51. Yellowfin shiner, *Notropis lutipinnis*. Spawning aggregation. William N. Roston. p. 95.

Pl. 52. Taillight shiner, *Notropis maculatus*. p. 95.

Pl. 53. Cape Fear shiner, *Notropis mekistocholas*. Richard G. Biggins, U.S. Fish and Wildlife Service. p. 97.

Pl. 54. Coastal shiner, *Notropis petersoni*. p. 95.

Pl. 55. Silver shiner, *Notropis photogenis*. p. 95.

Pl. 56. Swallowtail shiner, *Notropis procne*. p. 95.

Pl. 57. Rosyface shiner, *Notropis rubellus*. Richard T. Bryant. p. 96.

Pl. 58. Saffron shiner, *Notropis rubricroceus*. p. 95.

Pl. 59. New River shiner, *Notropis scabriceps*. p. 95.

Pl. 60. Sandbar shiner, *Notropis scepticus*. p. 95.

Pl. 61. Mirror shiner, *Notropis spectrunculus*. p. 95.

Pl. 62. Telescope shiner, *Notropis telescopus*. p. 96.

Pl. 63. Pugnose minnow, *Opsopoeodus emiliae*. Stephen T. Ross. p. 99.

Pl. 64. Fatlips minnow, *Phenacobius crassilabrum*. p. 100.

Pl. 65. Suckermouth minnow, *Phenacobius mirabilis*. Richard T. Bryant. p. 100.

Pl. 66. Kanawha minnow, *Phenacobius teretulus*. p. 100.

Pl. 67. Mountain redbelly dace, *Phoxinus oreas*. Male. p. 101.

Pl. 68. Mountain redbelly dace, *Phoxinus oreas*. Female. p. 101.

Pl. 69. Tennessee dace, *Phoxinus tennesseensis*. Male. Richard T. Bryant. p. 101.

Pl. 70. Bluntnose minnow, *Pimephales notatus*. Fred C. Rohde. p. 102.

Pl. 71. Fathead minnow, *Pimephales promelas*. Male. Richard T. Bryant. p. 102.

Pl. 72. Sailfin shiner, *Pteronotropis hypselopterus*. Male. p. 103.

Pl. 73. Blacknose dace, *Rhinichthys atratulus*. p. 103.

Pl. 74. Longnose dace, *Rhinichthys cataractae*. p. 103.

Pl. 75. Creek chub, *Semotilus atromaculatus*. p. 104.

Pl. 76. Fallfish, *Semotilus corporalis*. p. 104.

Pl. 77. Sandhills chub, *Semotilus lumbee*. Male. Fred C. Rohde. p. 104.

Pl. 78. Highfin carpsucker, *Carpiodes velifer*. Richard T. Bryant. p. 107.

Pl. 79. White sucker, *Catostomus commersoni*. p. 108.

Pl. 80. Creek chubsucker, *Erimyzon oblongus*. p. 109.

Pl. 81. Northern hog sucker, *Hypentelium nigricans*. p. 111.

Pl. 82. Roanoke hog sucker, *Hypentelium roanokense*. p. 111.

Pl. 83. Smallmouth buffalo, *Ictiobus bubalus*. p. 112.

Pl. 84. Spotted sucker, *Minytrema melanops*. p. 113.

Pl. 85. Black redhorse, *Moxostoma duquesnei*. p. 114.

Pl. 86. Golden redhorse, *Moxostoma erythrurum*. p. 114.

Pl. 87. Shorthead redhorse, *Moxostoma macrolepidotum*. p. 114.

Pl. 88. Bigeye jumprock, *Scartomyzon ariommus*. p. 116.

Pl. 89. Black jumprock, *Scartomyzon cervinus*. p. 116.

Pl. 90. Brassy jumprock, *Scartomyzon* new species. p. 117.

Pl. 91. Rustyside sucker, *Thoburnia hamiltoni*. p. 118.

Pl. 92. Flat bullhead, *Ameiurus platycephalus*. p. 119.

Pl. 93. White catfish, *Ameiurus catus*. p. 120.

Pl. 94. Yellow bullhead, *Ameiurus natalis*. p. 121.

Pl. 95. Brown bullhead, *Ameiurus nebulosus*. p. 122.

Pl. 96. Blue catfish, *Ictalurus furcatus*. p. 123.

Pl. 97. Channel catfish, *Ictalurus punctatus*. p. 123.

Pl. 98. Mountain madtom, *Noturus eleutherus*. p. 125.

Pl. 99. Carolina madtom, *Noturus furiosus*. p. 124.

Pl. 100. Orangefin madtom, *Noturus gilberti*. p. 124.

Pl. 101. Tadpole madtom, *Noturus gyrinus*. p. 124.

Pl. 102. Margined madtom, *Noturus insignis*. p. 124.

Pl. 103. Speckled madtom, *Noturus leptacanthus*. p. 124.

Pl. 104. Broadtail madtom, *Noturus* new species. Fred C. Rohde. p. 125.

Pl. 105. Flathead catfish, *Pylodictis olivaris*. p. 127.

Pl. 106. Redfin pickerel, *Esox americanus*. p. 128.

Pl. 107. Muskellunge, *Esox masquinongy*. Richard T. Bryant. p. 129.

Pl. 108. Chain pickerel, *Esox niger*. p. 130.

Pl. 109. Eastern mudminnow, *Umbra pygmaea*. p. 131.

Pl. 110. Rainbow trout, *Oncorhynchus mykiss*. p. 133.

Pl. 111. Brown trout, *Salmo trutta*. p. 134.

Pl. 112. Brook trout, *Salvelinus fontinalis*. Richard T. Bryant. p. 135.

Pl. 113. Pirate perch, *Aphredoderus sayanus*. p. 137.

Pl. 114. Swampfish, *Chologaster cornuta*. p. 139.

Pl. 115. Northern studfish, *Fundulus catenatus*. Male. J. R. Shute. p. 140.

Pl. 116. Golden topminnow, *Fundulus chrysotus*. Male. Fred C. Rohde. p. 141.

Pl. 117. Banded killifish, *Fundulus diaphanus*. Male. p. 142.

Pl. 118. Lined topminnow, *Fundulus lineolatus*. Male. Rudolf G. Arndt. p. 142.

Pl. 119. Speckled killifish, *Fundulus rathbuni*. p. 143.

Pl. 120. Waccamaw killifish, *Fundulus waccamensis*. Male (rear) and female. David G. Lindquist. p. 144.

Pl. 121. Bluefin killifish, *Lucania goodei*. Male. J. R. Shute. p. 144.

Pl. 122. Eastern mosquitofish, *Gambusia holbrooki*. Female. p. 146.

Pl. 123. Least killifish, *Heterandria formosa*. Female. Fred C. Rohde. p. 147.

Pl. 124. Sailfin molly, *Poecilia latipinna*. Female. p. 148.

Pl. 125. Brook silverside, *Labidesthes sicculus*. p. 150.

Pl. 126. Waccamaw silverside, *Menidia extensa*. p. 151.

Pl. 127. Fourspine stickleback, *Apeltes quadracus*. p. 153.

Pl. 128. Mottled sculpin, *Cottus bairdi*. p. 155.

Pl. 129. Banded sculpin, *Cottus carolinae*. p. 155.

Pl. 130. White perch, *Morone americana*. p. 157.

Pl. 131. White bass, *Morone chrysops*. Richard T. Bryant. p. 158.

Pl. 132. Striped bass, *Morone saxatilis*. p. 158.

Pl. 133. Mud sunfish, *Acantharchus pomotis*. p. 161.

Pl. 134. Rock bass, *Ambloplites rupestris*. p. 162.

Pl. 135. Flier, *Centrarchus macropterus*. p. 163.

Pl. 136. Blackbanded sunfish, *Enneacanthus chaetodon*. J. R. Shute. p. 164.

Pl. 137. Bluespotted sunfish, *Enneacanthus gloriosus*. J. R. Shute. p. 164.

Pl. 138. Banded sunfish, *Enneacanthus obesus*. p. 164.

Pl. 139. Redbreast sunfish, *Lepomis auritus*. p. 165.

Pl. 140. Green sunfish, *Lepomis cyanellus*. p. 166.

Pl. 141. Pumpkinseed, *Lepomis gibbosus*. p. 167.

Pl. 142. Warmouth, *Lepomis gulosus*. p. 167.

Pl. 143. Bluegill, *Lepomis macrochirus*. p. 168.

Pl. 144. Dollar sunfish, *Lepomis marginatus*. J. R. Shute. p. 169.

Pl. 145. Longear sunfish, *Lepomis megalotis*. Fred C. Rohde. p. 169.

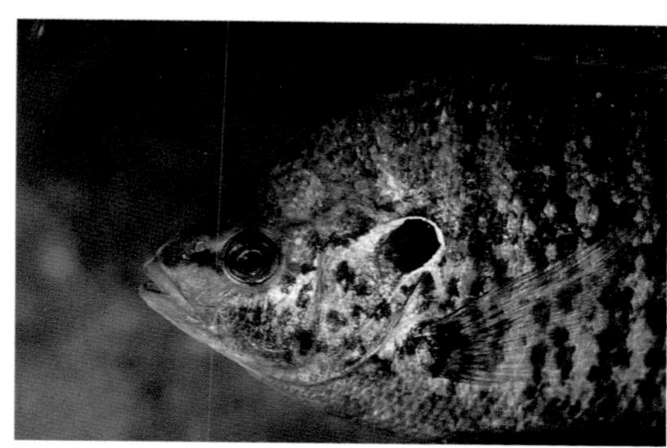

Pl. 146. Redear sunfish, *Lepomis microlophus*. p. 170.

Pl. 147. Spotted sunfish, *Lepomis punctatus*. p. 171.

Pl. 148. Redeye bass, *Micropterus coosae*. p. 171.

Pl. 149. Smallmouth bass, *Micropterus dolomieu*. p. 172.

Pl. 150. Spotted bass, *Micropterus punctulatus*. p. 172.

Pl. 151. Largemouth bass, *Micropterus salmoides*. p. 173.

Pl. 152. White crappie, *Pomoxis annularis*. Richard T. Bryant. p. 174.

Pl. 153. Black crappie, *Pomoxis nigromaculatus*. William N. Roston. p. 175.

Pl. 154. Carolina pygmy sunfish, *Elassoma boehlkei*. Male. p. 176.

Pl. 155. Everglades pygmy sunfish, *Elassoma evergladei*. Male on right, female on left. Fred C. Rohde. p. 176.

Pl. 156. Bluebarred pygmy sunfish, *Elassoma okatie*. Male. p. 176.

Pl. 157. Banded pygmy sunfish, *Elassoma zonatum*. Male. p. 176.

Pl. 158. Greenside darter, *Etheostoma blennioides*. Male. p. 179.

Pl. 159. Turquoise darter, *Etheostoma inscriptum*. Male. Fred C. Rohde. p. 179.

Pl. 160. Turquoise darter, *Etheostoma inscriptum*. Spawning pair, male on top. Steven R. Layman. p. 179.

Pl. 161. Kanawha darter, *Etheostoma kanawhae*. Male. p. 179.

Pl. 162. Candy darter, *Etheostoma osburni*. Male. Noel M. Burkhead. p. 179.

Pl. 163. Swannanoa darter, *Etheostoma swannanoa*. Male. p. 179.

Pl. 164. Seagreen darter, *Etheostoma thalassinum*. Male. p. 179.

Pl. 165. Banded darter, *Etheostoma zonale*. Male. p. 179.

Pl. 166. Johnny darter, *Etheostoma nigrum*. Female. p. 181.

Pl. 167. Tessellated darter, *Etheostoma olmstedi*. Male. p. 181.

Pl. 168. Waccamaw darter, *Etheostoma perlongum*. Female depositing eggs on underside of stick, male to the right. J. R. Shute. p. 181.

Pl. 169. Riverweed darter, *Etheostoma podostemone*. Male. p. 181.

Pl. 170. Glassy darter, *Etheostoma vitreum*. p. 183.

Pl. 171. Snubnose darter, *Etheostoma simoterum*. Male. p. 184.

Pl. 172. Blueside darter, *Etheostoma jessiae*. Male. p. 185.

Pl. 173. Sharphead darter, *Etheostoma acuticeps*. Male. p. 186.

Pl. 174. Sharphead darter, *Etheostoma acuticeps*. Female. p. 186.

Pl. 175. Greenfin darter, *Etheostoma chlorobranchium*. Male. p. 186.

Pl. 176. Greenfin darter, *Etheostoma chlorobranchium*. Female. p. 186.

Pl. 177. Redline darter, *Etheostoma rufilineatum*. Male. p. 186.

Pl. 178. Tippecanoe darter, *Etheostoma tippecanoe*. Male. Richard T. Bryant. p. 186.

Pl. 179. Wounded darter, *Etheostoma vulneratum*. Male. p. 186.

Pl. 180. Savannah darter, *Etheostoma fricksium*. Male. Fred C. Rohde. p. 187.

Pl. 181. Savannah darter, *Etheostoma fricksium*. Spawning pair, male on top. Steven R. Layman. p. 187.

Pl. 182. Pinewoods darter, *Etheostoma mariae*. Male. p. 187.

Pl. 183. Christmas darter, *Etheostoma hopkinsi*. Male. p. 188.

Pl. 184. Fantail darter, *Etheostoma flabellare*. Male p. 189.

Pl. 185. Carolina darter, *Etheostoma collis*. Female. p. 190.

Pl. 186. Swamp darter, *Etheostoma fusiforme*. Male. Fred C. Rohde. p. 190.

Pl. 187. Sawcheek darter, *Etheostoma serrifer*. p. 190.

Pl. 188. Yellow perch, *Perca flavescens*. p. 192.

Pl. 189. Blackbanded darter, *Percina nigrofasciata*. p. 192.

Pl. 190. Dusky darter, *Percina sciera*. Female. p. 192.

Pl. 191. Olive darter, *Percina squamata*. Male. p. 193.

Pl. 192. Piedmont darter, *Percina crassa*. Female. p. 194.

Pl. 193. Shield darter, *Percina peltata*. p. 194.

Pl. 194. Roanoke darter, *Percina roanoka*. Male. p. 194.

Pl. 195. Gilt darter, *Percina evides*. Male. p. 196.

Pl. 196. Tangerine darter, *Percina aurantiaca*. Male. William N. Roston. p. 197.

Pl. 197. Channel darter, *Percina copelandi*. Female. Noel M. Burkhead. p. 198.

Pl. 198. Blotchside logperch, *Percina burtoni*. p. 198.

Pl. 199. Logperch, *Percina caprodes*. p. 198.

Pl. 200. Roanoke logperch, *Percina rex*. Male. Noel M. Burkhead. p. 198.

Pl. 201. Sauger, *Stizostedion canadense*. Richard T. Bryant. p. 200.

Pl. 202. Walleye, *Stizostedion vitreum*. Richard T. Bryant. p. 201.

Pl. 203. Freshwater drum, *Aplodinotus grunniens*. Richard T. Bryant. p. 203.

many of the river mouths in the region were interconnected with fresh water, and many freshwater fishes then probably moved between river mouths and colonized new rivers. Generalized species were more likely to make such a trip than specialists.

Human beings, as a result of their present high level of technology and mobility, their huge numbers, and their penchant for tampering with things, are a major agent of introduction of fish species and have greatly accelerated the rate at which fishes move from one body of water to another. The effects of introductions by humans at some short time in the future will probably be seen as equal to the effects of continental drift, glaciers, and stream capture combined. While natural phenomena usually occur slowly, allowing fishes to adapt to and evolve with the changing conditions, the effects of human activity are occurring with explosive rapidity and over vast geographical areas, giving species but little time to adapt. The consequences will probably be a major shift in the composition of the world biota, including fishes, as we know it.

Collection and Study of Fishes

Many of the larger fishes of the mid-Atlantic region are of interest primarily to fishermen. There is neither space enough here to consider fishing techniques in detail for these numerous species nor need, for readers can refer to numerous readily available books and magazines on fishing. Data on record freshwater game fishes caught through 1991 in the region, by state, are presented, however, in Table 2.

There are millions of aquarium enthusiasts in the United States. Since most hobbyists buy their fishes from suppliers, they usually choose specimens from exotic and remote corners of the world. To learn about their charges they read books and articles, join hobbyist groups, call or write other hobbyists, and otherwise study where the fishes came from and how they live so that they can properly maintain them. Usually this approach works fairly well and specimens often live their allotted time on earth in relative comfort. The serious hobbyist may occasionally even make worthwhile biological contributions by recording observations and collecting data on species never before studied. Sometimes, in the case of the highly uninformed or very youthful hobbyist, fish may end their days prematurely. While unfortunate, such experiences are often filed away in the mind and may eventually help those persons develop into careful hobbyists or biologists.

While exotic—which usually means tropical—specimens have the advantage of at least partially satisfying our taste for things colorful and unfamiliar, they have several disadvantages. One is that they are often expensive. Another is that it can be difficult to learn adequately of their requirements, especially from firsthand experience. Third, an exotic that is not doing well and cannot be restored to vigorous health cannot be returned to its natural environment. Natives, however, are not as environmentally sensitive as tropicals since they are adapted to local conditions. Thus, natives should have no problem in coping with lower temperatures and reduced light, and they may even prefer them.

You say native fishes are not as colorful, as interesting, as challenging to maintain, as exotics. Not true! Many mid-Atlantic minnows, sunfishes, and darters are as colorful, as interesting (in shape, size, behavior, and habitat), and as challenging or easy to keep as many exotics. Such is the mystique of exotics that at least one local fish, the blackbanded sunfish, was of no interest to American suppliers until the mistaken belief that it was an exotic finally brought it to their attention.

How do you proceed to collect and raise native species? Before collecting any natives you will need to check with your state fish and game department to find out which might be endangered, threatened, or otherwise protected and what, if any, regulations govern the taking of unprotected species. A freshwater fishing license is required in many states, and there are usually limits on the types of gear and the sizes of nets that can be used to obtain specimens. Always be sure to get permission from any landowner before you attempt to collect.

Assuming that these possible hurdles have been cleared, you need to decide which nonprotected species you desire and how many specimens you want. Keep in mind

TABLE 2. *Sportfishing Records for Freshwater Game Fishes in the Mid-Atlantic Region through 1991 (weight in pounds/ounces)*

Species	S.C.	N.C.	Va.	Md.	Del.
Longnose gar	–	–	25/2	–	–
Bowfin	21/8	17/3	–	–	–
Hickory shad	–	2/14	–	4/0	–
American shad	7/0	7/15.5	–	8/2	6/12
Common carp	–	48/0	49/4	44/6	45/0
Smallmouth buffalo	–	61/0	–	–	–
White catfish	9/15	13/0	4/11	–	–
Bullhead catfish	6/3	–	–	–	–
Blue catfish	109/4	78/8	56/12	–	–
Channel catfish	58/0	40/8	28/12	–	18/0
Flathead catfish	74/0	62/7	56/0	–	–
Redfin pickerel	1/8.8	1/9	–	–	–
Northern pike	–	11/13	27/12	20/13	–
Muskellunge	–	38/0	45/0	–	–
Tiger muskellunge	–	33/8	–	26/12	–
Chain pickerel	6/4	8/0	7/1	6/8	7/3
Rainbow trout	9/6	16/5	12/9	14/3	–
Brown trout	17/9.5	15/13	18/4	11/9	9/4
Brook trout	2/5	7/7	5/10	4/12	–
White perch	1/13.5	1/14	2/1	2/10	2/4
White bass	4/13	5/14	6/13	–	–
Striped bass	55/0	54/2	42/6	–	–
Roanoke bass	–	2/8	2/6	–	–
Rock bass	–	1/4	2/2	–	–
Flier	1/4	1/5	–	–	–
Redbreast sunfish	2/0	1/12	–	–	–
Green sunfish	–	0/6	–	–	–
Pumpkinseed	1/0	–	–	–	–
Warmouth	2/2.5	1/13	–	–	–
Bluegill	3/4	4/5	–	3/0	2/1
Redear sunfish	3/7	4/6	4/12	2/5	–
Redeye bass	2/12	–	–	–	–
Smallmouth bass	6/12	10/2	7/7	8/4	4/15
Spotted bass	4/12	4/8	–	–	–
Largemouth bass	16/2	15/14	16/4	11/2	10/5
White crappie	5/1	–	–	–	–
Black crappie	5/0	–	–	–	–
Crappie	–	4/15	4/3	2/9	4/9
Yellow perch	3/4	2/9	2/2	2/3	2/11
Sauger	4/7	5/15	–	–	–
Walleye	9/3	13/8	12/15	11/4	–

the space you have available and thus the limitation on the number of specimens you can maintain. You then need to be able to find out in what part of your state, in which bodies of water, and in what habitats your desired specimens can be found.

Small, common species that live near the surface at the edges of quiet and shallow waters can be obtained easily from shore with an aquarium net or a sturdy, long-handled, fine-mesh dip net. To obtain larger species from more open water will require a seine, of which there are many types and sizes. Seines vary in material of construction (nylon, cotton, monofilament), length, height, mesh size, and whether they are flat or have a built-in bag to help restrain faster-moving fishes. Obviously, smaller and slower fishes in smaller bodies of water can be caught more readily with a smaller seine than larger, faster fishes that live in rivers or large lakes. Only experience will teach how best to find and catch fishes, just as confidence and dexterity in the handling of the net comes with experience. The more you read about the fishes you seek, the more wildlife films you see, and the more you watch fishes in the wild and in aquariums, the more you will learn about where and how to catch them. Do they rest mostly on the bottom? Do they swim actively in midwater or at the surface and as individuals or in a school? Do they hide during the day and become active at night? The answers to these questions indicate behaviors that will give you clues on where and how to look for the fishes you want.

Fish nets (dip nets, seines) are extremely simple in concept yet still the most fool-proof, cheap, and easy-to-transport tools for collecting fishes. They were invented and used in Stone Age societies many thousands of years ago. A nylon seine measuring ten feet by four feet with one-eighth-inch mesh can be used in most habitats. It is best used with a partner, but it is small enough, especially when partly rolled up, so that it can be used when working alone. Seining in shallow open water is easily understood and done, even by the novice. But what do you do in more difficult situations?

In swift water, and with the help of others, hold the net stationary, push it into soft substrate, or wedge it against rocks, trees, or other stationary objects; kick the substrate vigorously upstream to flush fishes toward the net; and then raise the net. (See Fig. 14.) In deep water next to the shore, reach the poles out away from the bank, make a large scoop down into the water, trapping the fishes against the bank, and then lift the net up while holding it against the bank. When underwater obstructions are present, partly roll up each end of the net on the poles and make a giant scoop with the remainder. Dip it against, or partly around, an obstruction and then deftly lift it.

Seining is best done by pulling against the current. This way, in order to escape, fishes in the net have to swim against the current while they are "off-balance" and pushed against the seine. Only the larger and stronger fishes might escape. However, some schooling fishes that typically remain stationary or swim slowly into a current are best caught by seining quickly with the current. This technique may also be profitably used where the current is so strong that seining against it is impossible. The added speed derived by seining with a swift current, and a bit of luck, can even result in the capture of larger and swifter fishes.

For bottom dwellers, make sure that the bottom of the seine is tight against the substrate. Disturbance of rocks, gravel, mud, leaves, and twigs on the bottom by kicking and splashing is important to "flush out" fishes. This roiling of the bottom

Figure 14. Kick seining. David G. Lindquist.

and the resultant increase in turbidity also helps to confuse and blind them. The young of some fishes, such as lampreys, live in soft substrate and are caught only by actively dislodging them from mud. Simply pulling a seine over the bottom is not sufficient, as they will be missed. Remember, however, to avoid excessive and unnecessary disturbance of the habitat.

Make several short runs with the seine and then check the contents. This is usually more productive than making a smaller number of long runs and will also hedge against the possibility that the net will catch on a rock or branch, allowing most or all of the contained fishes to escape.

At the end of a seine haul it is usually best to slide the seine smoothly onto a beach or sandbar rather than to lift it out of the water while away from shore, as this will also catch fishes that usually swim ahead of the net but are not yet in it. However, when you are seining at some distance away from shore, the net contents must be examined while still out in the water. In this case, slide the net smoothly, evenly, and quickly out of the water. Again, adapt how you use the seine to local conditions and to the species sought.

When moving between sites, clean the seine and shake it dry so as not to introduce animals or plants from one area to another. When seining is completed for the day, clean and shake the net, dry it thoroughly, and then roll it up on the poles. This way you will have a clean net ready for your next trip, and you will prevent net rot.

It is best to take only a few specimens of the species you desire. Taking a larger number often mitigates against successful transport to, and maintenance of, the specimens in aquariums, and it can have an adverse impact on the wild populations.

Treat all fishes as delicately as possible. Avoid touching them; move them with a dip net or pour them gently. And when you must touch them, first wet your hands.

When transporting and subsequently housing fishes, make sure that they are not crowded, that the temperature is approximately the one at which they were caught, that they receive adequate oxygen, and that other water quality characteristics (pH, hardness, color, etc.) are within the range of tolerance of the species. Taking water from the site of capture will help achieve proper water chemistry.

Before collecting any fishes, you should read about them. It is beyond the scope of this book to consider details of aquarium size, filtration, lighting, etc., but such information is readily available elsewhere. The beginner is advised to start with easy-to-obtain, easy-to-keep species, and with tried and conventional techniques. Generally speaking, the easier-to-obtain fishes are those that will do best in aquariums since they are adapted to a wide range of conditions or to widely prevailing conditions. Once you have developed a "feel" for your charges, intelligent experimentation may well lead to improvements over current dogma. No doubt you will make mistakes, but if these are not too serious, you can consider yourself as doing well. Sometimes there is no substitute for failing and then trying new things.

There is still much to be learned about the fishes of the mid-Atlantic region (as well as other regions), and the hobbyist can make worthwhile observations. Observation is difficult work, however, and it requires extensive preliminary library and other background research followed by a careful and thorough approach in order to record details with pen, film, and tape recorder. Details of behavior are needed for many species, as are other aspects of life history. For many species, for example, we lack adequate information on the number of eggs or young produced, their size, how often they are produced, whether they sink or float, the time that elapses before they hatch, the food they need, their rate of growth, and the time it takes them to reach maturity. Consult more advanced friends or colleagues, books, magazines, and experts at schools, aquariums, museums, zoos, and pet stores when you have questions.

The authors also encourage photography of fishes. This is a fun and instructive hobby that can lead to great personal satisfaction and new discoveries, as well as making photographs available for use in magazines and books. Photography can be in controlled situations (aquariums) or in the wild. In the latter case, photographs of the habitat are also highly desirable, especially for poorly known fishes.

There are many magazines and books available on fish photography, as you will see by visiting your library or bookstore. A 35-mm camera with a lens that can focus from infinity to close-up and fast film will allow photographs in natural light. Indoor work requires a flash or strobe light. Fishes can be photographed in aquariums in the home or in small aquariums or special "photochambers" in the field; all of these can be purchased or can be built of cut and glued glass or plexiglass. Glass is much harder to scratch than plexiglass, but plexiglass is less breakable and therefore less liable to result in cuts.

Build the chamber to accommodate the size of the fishes. The idea behind chamber construction is to facilitate the handling of fishes and to have a natural substrate in order to create a natural-appearing and aesthetically pleasing photograph. The chambers are usually only a fraction of an inch to a few inches wide, in order to restrict the fish to a narrow plane of movement that facilitates photography, but constructing even the smallest chamber suitable for a given fish requires moving a minimal amount of water and substrate.

Photographs of mating, nest building (see Fig. 15), egg deposition (see Fig. 16),

Figure 15. Nests of pumpkinseed, *Lepomis gibbosus*. Rudolf G. Arndt.

Figure 16. Egg cluster of madtom, *Noturus* species. David G. Lindquist.

care of young, etc., usually require working with a larger, long-term indoor setup. This is usually for the more advanced hobbyist, who will know the relevant literature and equipment.

Most of the photographs in this book were taken of fishes in aquariums with a 35-mm single-lens, through-the-lens-focusing reflex camera with a 105-mm macro lens. This allows a close focus for most small fishes. Strobes automatically adjust the proper light output. The key, however, is to polarize both the lens and the flash output and to keep the polarizing filters at right angles to one another. This usually eliminates reflections.

Thoroughly clean both the front and the back of each sheet of glass of an aquarium and equip the tank with a good filtering system. Supply the aquarium with a substrate of natural materials from the natural habitat of the fishes to be photographed. Place a dark and soft background, such as black velvet, behind the aquarium. Obviously, larger species will require larger aquariums and larger flash units. Such techniques and equipment have been used for many years with marine systems.

Observation and photography of fishes in the wild is a growing and challenging sport and scientific endeavor, spurred partly by the increased technology in easily portable diving and photography equipment. We greatly encourage such field work. Underwater study and photography of fishes, especially in clear tropical salt waters, is now decades old and was fostered especially by scuba equipment and the riot of colors on coral reefs. A newer activity is snorkeling in fresh waters, especially larger creeks and rivers. The minimum equipment is a face mask and snorkel, although swim fins are also helpful. A wet suit, underwater light, plastic slate and pencil, camera and strobe light, collecting net, and plastic bag will prepare you for just about any weather and will enable you to photograph, collect, and bring back samples. Scuba gear can be of great assistance, although it is relatively expensive. Snorkeling or diving at night will often give a very different picture of fish species composition and abundance from that observed during the day. Also, many species can be approached more readily at night than during the day. As always, making detailed notes and photographs is of extreme importance. Of course, you should be healthy and a good swimmer, avoid stormy weather, always go with a partner, and know the waters in which you plan to dive.

Remember that a good biologist and hobbyist is always conservation-minded. Simply put, you take only what you need (of those organisms whose capture is not prohibited) and only what you can properly care for. If your collection, maintenance, and photography has a purpose beyond confinement and simple observation, such as an attempt to learn and document something new about fishes, so much the better. Ask specific questions about your subjects and try to answer them. This will lead to more rapid learning than just looking at them with no specific questions in mind.

If you should tire of your hobby, or of certain locally caught specimens, and if they are healthy, it is reasonable to release them. However, you should realize that the distribution of many species is but imperfectly known. A release in a habitat that seems likely but is not in fact one in which a given species occurs may lead to erroneous conclusions about the fish's distribution and abundance if your released specimen is later caught by a biologist. Further, biologists today have such sophisticated methods of study that they can often recognize local populations, called demes, which are often genetically but not morphologically (i.e., in physical appearance) different from other populations of a species. The evolution of demes is a current area of study by many scientists, and the release of specimens at areas other than those where they were taken can cause problems for scientific observation and study. Also, numerous introduced species have become, or threaten to become, pests. You could cause serious environmental repercussions through an inappropriate release. In many cases releases are illegal. Specimens of unknown origin and specimens that cannot feasibly be released where they were caught are best maintained until they die or given to another person who will maintain them.

Make sure you keep accurate records of where specimens were caught and release them only there. Captive spawnings should be undertaken only with parents from a specific population and the young released only where the parents were taken.

Make sure that conditions for release are appropriate. For example, don't release in mid-winter fishes that were caught last June and kept in a heated living room, as they almost certainly will not survive the shock of the change in temperature.

If you have questions, consult an expert at a college, university, museum, zoo or other appropriate institution for help. Good luck.

Some Impacts of Humans on the Fishes of the Region

Wherever people go they change the environment. This fact has had a major and irrevocable impact on the earth and its biota. The fish fauna of the mid-Atlantic region has not been exempt, and some of the changes affecting it are mentioned briefly here, primarily to suggest caution in the release and introduction of fishes. See the species accounts later in the book for more detail on most of the species mentioned below.

The goldfish is a native of Eurasia. It has long been a popular aquarium fish and has been selected for a number of unnatural colors and shapes by man. Through accidental and intentional releases, it now can be found in quiet and warm waters throughout the mid-Atlantic region. The common carp is also a native of Eurasia, where it is often raised as a food fish because it is adapted for life in ponds and other quiet waters that are readily available or can be easily duplicated on farms. It was brought to this country in the 1800s to be raised for food. Escapes and releases have resulted in its becoming widely established in the region.

The grass carp is native to eastern Asia. It was brought into this country recently because of its ability to eat huge quantities of aquatic vegetation and is used to clean out waters suffering from eutrophication, typically caused by human activities. In order to prevent rampant colonization by this fish, nonreproductive strains of grass carp are now being introduced. Nevertheless, some of the earlier strains have already invaded numerous areas of the United States, and the species threatens to become a pest.

The brown trout, a large, feisty, tasty game fish native to Europe, Asia, and Africa, was long ago introduced into the United States as a game species. It is now found throughout the parts of the mid-Atlantic region where higher elevations and colder waters prevail. It apparently coexists well with native species.

At least six species of fishes native to other portions of the United States but not to the mid-Atlantic region have been introduced here and are now established: threadfin shad, red shiner, blue catfish, northern pike, rainbow trout, and sockeye salmon. These species are native to waters further west, except for the northern pike, which is native to cooler waters to the north. Most were introduced as game or food fishes, although the threadfin shad was introduced as a forage fish and the red shiner was introduced accidentally through bait or aquarium specimen releases.

In addition, many of the fishes in the region, such as the channel catfish, blue catfish, flathead catfish, green sunfish, largemouth bass, and white crappie, were native to only small parts of it but as a result of introductions now occur throughout.

Four species that once occurred in the region have, or are believed to have, disappeared (been extirpated) here. One of these is the lake sturgeon, *Acipenser fulvescens*, known in the region from eight specimens taken in 1945 in the French Broad River, a tributary of the Tennessee River, in North Carolina. It is listed as being of special concern in North Carolina. The trout-perch, *Percopsis omiscomaycus*, a small nongame species common in waters north and west of the region, was formerly found in

TABLE 3. *Current Federal and State Legal Status, and Proposed Status,*[a] *of Freshwater Fishes of the Carolinas, Virginia, Maryland, and Delaware*

	U.S.	S.C.	N.C.	Va.	Md.	Del.
Lampreys						
Least brook lamprey	–	–	S	–	–	–
American brook lamprey	–	–	T	–	–	–
Sturgeons						
Shortnose sturgeon	E	E	E	E	E	E
Atlantic sturgeon	–	S	S	–	–	–
Lake sturgeon	–	–	S*	–	–	–
Paddlefishes						
Paddlefish	–	–	E	T	–	–
Mooneyes						
Mooneye	–	–	S	–	–	–
Herrings						
Blueback herring (Santee-Cooper)	–	S	–	–	–	–
Hickory shad	–	S	–	–	–	–
Carps and minnows						
Rosyside dace (subspecies)	–	–	S	–	–	–
Whitetail shiner	–	PS	–	–	–	–
Spotfin chub	T	–	T	T	–	–
Steelcolor shiner	–	–	–	T	–	–
Slender chub	T	–	–	T	–	–
Cutlips minnow	–	–	E	–	–	–
Rosyface chub	–	–	T	–	–	–
Santee chub (coastal plain)	–	PS	S	–	–	–
Striped shiner	–	–	T	–	–	–
Warpaint shiner	–	PS	–	–	–	–
River chub	–	PS	–	–	–	–
Whitemouth shiner	–	–	–	T	–	–
Emerald shiner	–	–	–	T	–	–
Bridle shiner	–	PS	S	–	–	–
Redlip shiner	–	PS	–	–	–	–
Tennessee shiner	–	PS	–	–	–	–
Yellowfin shiner	–	–	S	–	–	–
Cape Fear shiner	E	–	E	–	–	–
Mirror shiner	–	PS	–	–	–	–
Kanawha minnow	–	–	S	–	–	–
Tennessee dace	–	–	–	E	–	–
Blacknose dace	–	PS	–	–	–	–
Cheat minnow	–	–	–	–	E	–
Longnose dace	–	PS	–	–	–	–
Sandhills chub	–	PS	S	–	–	–

TABLE 3. *Continued*

	U.S.	S.C.	N.C.	Va.	Md.	Del.
Suckers						
River carpsucker	–	–	S	–	–	–
Highfin carpsucker	–	–	S	–	–	–
Longnose sucker	–	–	–	–	E	–
Bigeye jumprock	–	–	S	–	–	–
River redhorse	–	PS	S	–	–	–
V-lip redhorse	–	PS	–	–	–	–
Rustyside sucker	–	–	E	–	–	–
Bullhead catfishes						
Mountain madtom	–	–	S	–	–	–
Yellowfin madtom	T	–	–	T	–	–
Stonecat	–	–	E	–	–	–
Carolina madtom	–	–	S	–	–	–
Orangefin madtom	–	–	E	T	–	–
Broadtail madtom	–	PT	S	–	–	–
Trouts						
Brook trout	–	S	–	–	–	–
Trout-perches						
Trout-perch	–	–	–	–	E*	–
Topminnows and killifishes						
Waccamaw killifish	–	–	S	–	–	–
Bluefin killifish	–	–	S	–	–	–
Livebearers						
Least killifish	–	–	S	–	–	–
Silversides						
Waccamaw silverside	T	–	T	–	–	–
Sculpins						
Banded sculpin	–	–	T	–	–	–
Slimy sculpin	–	–	–	–	T	–
Sunfishes						
Rock bass	–	PS	–	–	–	–
Blackbanded sunfish	–	–	–	E	IN	–
Pygmy sunfishes						
Carolina pygmy sunfish	–	PT	T	–	–	–
Bluebarred pygmy sunfish	–	S	–	–	–	–
Perches						
Western sand darter	–	–	–	T	–	–
Sharphead darter	–	–	T	E	–	–
Greenfin darter	–	–	–	T	–	–
Carolina darter	–	PS	S	T	–	–
Fantail darter	–	PS	–	–	–	–
Turquoise darter	–	–	S	–	–	–
Blueside darter	–	–	S*	–	–	–

TABLE 3. *Continued*

	U.S.	S.C.	N.C.	Va.	Md.	Del.
Pinewoods darter	–	PE	S	–	–	–
Waccamaw darter	–	–	T	–	–	–
Riverweed darter	–	–	S	–	–	–
Maryland darter	E	–	–	–	E	–
Snubnose darter	–	–	S*	–	–	–
Tippecanoe darter	–	–	–	T	–	–
Variegate darter	–	–	–	E	–	–
Glassy darter	–	–	–	–	E	–
Wounded darter	–	–	S	–	–	–
Banded darter	–	PT	–	–	–	–
Duskytail darter	–	–	–	E	–	–
Blotchside logperch	–	–	E	–	–	–
Logperch	–	–	T	–	–	–
Longhead darter	–	–	–	T	–	–
Stripeback darter	–	–	–	–	E	–
Sharpnose darter	–	–	S	–	–	–
Roanoke logperch	E	–	–	E	–	–
Dusky darter	–	–	E	–	–	–
Olive darter	–	–	S	–	–	–
Drums						
Freshwater drum	–	–	T	–	–	–

a. E = endangered, T = threatened, S = special concern, IN = in need of conservation, PE = proposed endangered, PT = proposed threatened, PS = proposed special concern. An * indicates that the species is probably already extirpated.

the Potomac River drainage of northern Virginia and Maryland. It is officially listed as extirpated by Maryland. The blackside darter, *Percina maculata*, has apparently been extirpated from the mountains of Virginia, the only portion of the mid-Atlantic region where it occurred. It is still, however, widely distributed in much of the central United States and adjacent Canada. The ashy darter, *Etheostoma cinereum*, occurred within the region only in extreme western Virginia, and it was last noted there in 1964; it was taken in the Clinch River, Tennessee, just below the Virginia state line in 1992. It is rare in most of the rest of its range in Tennessee and portions of adjacent states.

The harelip sucker, *Lagochila lacera*, was formerly found in the mid-Atlantic region only in western Virginia, in the North Fork Holston River, where the first and only specimens reported in the region were seen in 1888. Named for its peculiar harelike lips, this sucker reached a length of about eighteen inches and a weight of two or three pounds, and it once was a valuable food fish. It occurred in rivers and fed on small mollusks such as snails and limpets. It was formerly found in much of the central United States, but the species probably became extinct in the early 1900s.

Land clearing and poor agricultural practices no doubt resulted in excessive water runoff and siltation that smothered its food and hindered its feeding.

A number of other fishes in the region are today of special concern (see Table 3), a classification denoting species that need special attention to safeguard their existence. Some occur only marginally in the region but are still doing well in other portions of their range, some are doing poorly throughout their range, and some are in need of further study before we can determine their actual status.

Common Names and Scientific Names

While most of us know a particular species of fish by a common name, such names vary widely and are of limited value as a communication tool. For example, the bluegill is known as bream or pond perch by some, the white crappie and black crappie are both often known just as crappie, and the sauger is sometimes known as jack salmon, spotted jack, and sand pike. In an effort to increase communication, especially among nonscientists, common names of many groups of animals and plants have been standardized, including all the fishes found in the United States. Such lists have been published and are generally available.

Despite standardization of common names, each extant and extinct species of life has been given a scientific name so that scientists and others can identify it and relate it to its relatives. The application and use of a scientific name must follow long and well-established rules too complicated and detailed to explain here; such names are changed only rarely and only after complex rules are followed.

The earliest scientific names still in use today date to 1758, when Karl Linnaeus, in the tenth edition of his famous work, *Systema Naturae* (*System of Nature*), was the first person consistently to employ what is now known as the binomial system of nomenclature, in which each species is given two names. The first name is the genus name and the second is the species name. The first letter of the genus name is always capitalized, while the species name for fishes and other animals begins with a lowercase letter, and both names are always underlined or italicized to indicate that they are scientific names. Linnaeus, a Swedish botanist, was a prodigious worker who solicited animals and plants from all over the world and then described and named them. He named hundreds of animals and plants, including the common carp, goldfish, white catfish, pumpkinseed, and redbreast sunfish.

Scientific names seem foreign and cumbersome to most people, because they are based on Latin. If one knows Latin, however, the names are highly understandable and usually descriptive of something characteristic or unique about a given species, such as its size, color, shape, behavior, habitat, or distribution. For example, the genus name for the pygmy sunfishes *Elassoma* means "small," the species name *punctatus* for another fish means "spotted," and the species name *kanawhae* refers to the Kanawha River, where another species was first found or now occurs. Each of these names tells us something about the named species. Some other species have names that honor a researcher, a patron who supported research on it, or the person who first collected that species. For example, the species name *boehlkei* is named in honor of ichthyologist James E. Böhlke. Sometimes the name may be merely whimsical and have no apparent significance, such as the genus names *Lucania* and *Morone*.

In formal treatises the scientific name is often followed by a nonitalicized name or names, which is/are the name(s) of the author(s) or person(s) who first described that species and published the name and description in a generally available publication. Some species have, for a variety of reasons, a number of synonyms or scientific names by which they were known in the past that were changed for compelling reasons. Thus a species might be referred to by several names over time. See the introduction to the species accounts for an explanation of the naming scheme utilized in this book.

Freshwater Fishes of the Region

INTRODUCTION

The following section provides information on the 262 species of fishes known to occur in South Carolina, North Carolina, Virginia, Maryland, and Delaware and considered by the authors to be freshwater species. Included are those fishes that spend all of their life in fresh water, those that move into salt water for most of their life after beginning life in fresh water (anadromous fishes), those that move into fresh water for most of their life after beginning life in salt water (catadromous), and those that are primarily found in salt water or brackish water but that frequently are found in fresh water, typically to find food. Both natives and established exotics are covered, and unsuccessful introductions are mentioned. A checklist of all species of freshwater fishes in the region is given in Table 4.

Important sport and commercial species and the more common but less well known smaller fishes are discussed in individual accounts. A single account is provided in cases where several highly similar smaller fishes occur within a genus.

Individual accounts include both the common and the scientific name. The common name used is the one chosen by a group of scientists and published in an effort to increase effective communication. Until recently, the standardization of names was achieved only with scientific names, but these are not in common use by nonscientists, which is one reason the standardized common names are given here. Scientific names still provide the ultimate basis for naming and identifying a particular species, however, and the serious student will benefit from learning them. The authors' use of common names and the systematic sequence of families and classes follow those in C. R. Robins et al., *Common and Scientific Names of Fishes* (1991), with the exceptions of the use here of the names Moronidae and Elassomatidae. Scientific names of species follow those given in L. M. Page and B. M. Burr, *A Field Guide to Freshwater Fishes* (1991).

Photographs are provided for the most commonly encountered species. Where there are several similar species in a genus, a single representative is illustrated. Photographs are not provided for some rare and localized fishes.

A brief description, which includes the size range (in total length) of adults in both English and metric units, is provided for each entity. Where the account covers more than one species, the size range given includes all species considered. While we would have liked to provide a full account for each fish species in the region, this would not have been practical. As a compromise, we mention each species, but for less common ones we often do so only briefly and in the account of a more common species. The names of species that are mentioned in this way appear in boldface within the text for ease of reference. (Of course *all* species mentioned are listed by both their common and scientific names in the index.) Weights are provided for the larger game species. Table 2 provides record sizes (through 1991) for sport fishes caught in each state of the region.

TABLE 4. *Freshwater Fishes of the Carolinas, Virginia, Maryland, and Delaware and Their State(s) of Occurrence*

Common name	Scientific name	S.C.	N.C.	Va.	Md.	Del.
Lampreys	Petromyzontidae					
Ohio lamprey	*Ichthyomyzon bdellium*	–	–	x	–	–
Mountain brook lamprey	*I. greeleyi*	–	x	x	–	–
Least brook lamprey	*Lampetra aepyptera*	–	x	x	x	x
American brook lamprey	*L. appendix*	–	x	x	x	x
Sea lamprey	*Petromyzon marinus*	x	x	x	x	x
Sturgeons	Acipenseridae					
Shortnose sturgeon	*Acipenser brevirostrum*	x	x	x	x	x
Atlantic sturgeon	*A. oxyrhynchus*	x	x	x	x	x
Paddlefishes	Polyodontidae					
Paddlefish	*Polyodon spathula*	–	x	x	–	–
Gars	Lepisosteidae					
Longnose gar	*Lepisosteus osseus*	x	x	x	x	x
Bowfins	Amiidae					
Bowfin	*Amia calva*	x	x	x	x	–
Mooneyes	Hiodontidae					
Mooneye	*Hiodon tergisus*	–	x	–	–	–
Freshwater eels	Anguillidae					
American eel	*Anguilla rostrata*	x	x	x	x	x
Herrings	Clupeidae					
Blueback herring	*Alosa aestivalis*	x	x	x	x	x
Hickory shad	*A. mediocris*	x	x	x	x	x
Alewife	*A. pseudoharengus*	x	x	x	x	x
American shad	*A. sapidissima*	x	x	x	x	x
Gizzard shad	*Dorosoma cepedianum*	x	x	x	x	x
Threadfin shad	*D. petenense*	x	x	x	x	x
Carps and minnows	Cyprinidae					
Central stoneroller	*Campostoma anomalum*	x	x	x	x	–
Goldfish	*Carassius auratus*	x	x	x	x	x
Redside dace	*Clinostomus elongatus*	–	–	–	x	–
Rosyside dace	*C. funduloides*	x	x	x	x	x
Grass carp	*Ctenopharyngodon idella*	x	x	x	x	x
Satinfin shiner	*Cyprinella analostana*	x	x	x	x	x
Greenfin shiner	*C. chloristia*	x	x	–	–	–
Whitetail shiner	*C. galactura*	x	x	x	–	–
Bannerfin shiner	*C. leedsi*	x	–	–	–	–
Red shiner	*C. lutrensis*	–	x	–	–	–
Spotfin chub	*C. monacha*	–	x	x	–	–
Whitefin shiner	*C. nivea*	x	x	–	–	–
Fieryblack shiner	*C. pyrrhomelas*	x	x	–	–	–
Spotfin shiner	*C. spiloptera*	–	x	x	x	x
Steelcolor shiner	*C. whipplei*	–	–	x	–	–

TABLE 4. *Continued*

Common name	Scientific name	S.C.	N.C.	Va.	Md.	Del.
Common carp	*Cyprinus carpio*	x	x	x	x	x
Silverjaw minnow	*Ericymba buccata*	–	–	x	x	–
Slender chub	*Erimystax cahni*	–	–	x	–	–
Streamline chub	*E. dissimilis*	–	–	x	–	–
Blotched chub	*E. insignis*	–	x	x	–	–
Tonguetied minnow	*Exoglossum laurae*	–	x	x	–	–
Cutlips minnow	*E. maxillingua*	–	x	x	x	x
Eastern silvery minnow	*Hybognathus regius*	x	x	x	x	x
Bigeye chub	*Hybopsis amblops*	–	x	x	–	–
Highback chub	*H. hypsinotus*	x	x	x	–	–
Thicklip chub	*H. labrosa*	x	x	x	–	–
Rosyface chub	*H. rubrifrons*	x	x	–	–	–
Santee chub	*H. zanema*	x	x	–	–	–
White shiner	*Luxilus albeolus*	–	x	x	–	–
Crescent shiner	*L. cerasinus*	–	x	x	–	–
Striped shiner	*L. chrysocephalus*	–	x	x	–	–
Warpaint shiner	*L. coccogenis*	x	x	x	–	–
Common shiner	*L. cornutus*	–	–	x	x	x
Rosefin shiner	*Lythrurus ardens*	–	x	x	–	–
Mountain shiner	*L. lirus*	–	–	x	–	–
Pinewoods shiner	*L. matutinus*	–	x	–	–	–
Pearl dace	*Margariscus margarita*	–	–	x	x	–
Bluehead chub	*Nocomis leptocephalus*	x	x	x	–	–
River chub	*N. micropogon*	x	x	x	x	x
Bigmouth chub	*N. platyrhynchus*	–	x	x	–	–
Bull chub	*N. raneyi*	–	x	x	–	–
Golden shiner	*Notemigonus crysoleucas*	x	x	x	x	x
Whitemouth shiner	*Notropis alborus*	x	x	x	–	–
Highfin shiner	*N. altipinnis*	x	x	x	–	–
Comely shiner	*N. amoenus*	x	x	x	x	x
Popeye shiner	*N. ariommus*	–	–	x	–	–
Emerald shiner	*N. atherinoides*	–	–	x	–	–
Bridle shiner	*N. bifrenatus*	x	x	x	x	x
Ironcolor shiner	*N. chalybaeus*	x	x	x	x	x
Redlip shiner	*N. chiliticus*	x	x	x	–	–
Greenhead shiner	*N. chlorocephalus*	x	x	–	–	–
Dusky shiner	*N. cummingsae*	x	x	–	–	–
Spottail shiner	*N. hudsonius*	x	x	x	x	x
Tennessee shiner	*N. leuciodus*	x	x	x	–	–
Sand shiner	*N. ludibundus*	–	–	x	–	–
Yellowfin shiner	*N. lutipinnis*	x	x	–	–	–
Taillight shiner	*N. maculatus*	x	x	–	–	–
Cape Fear shiner	*N. mekistocholas*	–	x	–	–	–

TABLE 4. *Continued*

Common name	Scientific name	S.C.	N.C.	Va.	Md.	Del.
Coastal shiner	*N. petersoni*	x	x	–	–	–
Silver shiner	*N. photogenis*	–	x	x	–	–
Swallowtail shiner	*N. procne*	x	x	x	x	x
Rosyface shiner	*N. rubellus*	–	x	x	x	x
Saffron shiner	*N. rubricroceus*	–	x	x	–	–
New River shiner	*N. scabriceps*	–	x	x	–	–
Sandbar shiner	*N. scepticus*	x	x	–	–	–
Roughhead shiner	*N. semperasper*	–	–	x	–	–
Mirror shiner	*N. spectrunculus*	x	x	x	–	–
Telescope shiner	*N. telescopus*	–	x	x	–	–
Mimic shiner	*N. volucellus*	–	x	x	–	–
Sawfin shiner	*Notropis* new species	–	–	x	–	–
Pugnose minnow	*Opsopoeodus emiliae*	x	–	–	–	–
Fatlips minnow	*Phenacobius crassilabrum*	–	x	x	–	–
Suckermouth minnow	*P. mirabilis*	–	–	x	–	–
Kanawha minnow	*P. teretulus*	–	x	x	–	–
Stargazing minnow	*P. uranops*	–	–	x	–	–
Mountain redbelly dace	*Phoxinus oreas*	–	x	x	–	–
Tennessee dace	*P. tennesseensis*	–	–	x	–	–
Bluntnose minnow	*Pimephales notatus*	–	x	x	x	–
Fathead minnow	*P. promelas*	–	x	x	x	x
Bullhead minnow	*P. vigilax*	–	–	x	–	–
Sailfin shiner	*Pteronotropis hypselopterus*	x	–	–	–	–
Blacknose dace	*Rhinichthys atratulus*	x	x	x	x	x
Cheat minnow	*R. bowersi*	–	–	–	x	–
Longnose dace	*R. cataractae*	x	x	x	x	x
Creek chub	*Semotilus atromaculatus*	x	x	x	x	x
Fallfish	*S. corporalis*	–	–	x	x	x
Sandhills chub	*S. lumbee*	x	x	–	–	–
Suckers	Catostomidae					
River carpsucker	*Carpiodes carpio*	–	x	–	–	–
Quillback	*C. cyprinus*	x	x	x	x	–
Highfin carpsucker	*C. velifer*	x	x	–	–	–
Longnose sucker	*Catostomus catostomus*	–	–	–	x	–
White sucker	*C. commersoni*	x	x	x	x	x
Creek chubsucker	*Erimyzon oblongus*	x	x	x	x	x
Lake chubsucker	*E. sucetta*	x	x	x	–	–
Northern hog sucker	*Hypentelium nigricans*	x	x	x	x	x
Roanoke hog sucker	*H. roanokense*	–	x	x	–	–
Smallmouth buffalo	*Ictiobus bubalus*	x	x	–	–	–
Bigmouth buffalo	*I. cyprinellus*	–	x	–	–	–
Spotted sucker	*Minytrema melanops*	x	x	–	–	–
Silver redhorse	*Moxostoma anisurum*	x	x	x	–	–

TABLE 4. *Continued*

Common name	Scientific name	S.C.	N.C.	Va.	Md.	Del.
River redhorse	*M. carinatum*	–	x	x	–	–
Black redhorse	*M. duquesnei*	–	x	x	–	–
Golden redhorse	*M. erythrurum*	–	x	x	x	–
Shorthead redhorse	*M. macrolepidotum*	x	x	x	x	x
V-lip redhorse	*M. pappillosum*	x	x	x	–	–
Robust redhorse	*M. robustum*	x	x	–	–	–
Bigeye jumprock	*Scartomyzon ariommus*	–	x	x	–	–
Black jumprock	*S. cervinus*	–	x	x	–	–
Striped jumprock	*S. rupiscartes*	x	x	–	–	–
Brassy jumprock	*Scartomyzon* new species	x	x	x	–	–
Rustyside sucker	*Thoburnia hamiltoni*	–	x	x	–	–
Torrent sucker	*T. rhothoeca*	–	–	x	–	–
Bullhead catfishes	Ictaluridae					
Snail bullhead	*Ameiurus brunneus*	x	x	x	–	–
White catfish	*A. catus*	x	x	x	x	x
Black bullhead	*A. melas*	x	x	x	–	–
Yellow bullhead	*A. natalis*	x	x	x	x	x
Brown bullhead	*A. nebulosus*	x	x	x	x	x
Flat bullhead	*A. platycephalus*	x	x	x	–	–
Blue catfish	*Ictalurus furcatus*	x	x	x	–	–
Channel catfish	*I. punctatus*	x	x	x	x	x
Mountain madtom	*Noturus eleutherus*	–	x	x	–	–
Yellowfin madtom	*N. flavipinnis*	–	–	x	–	–
Stonecat	*N. flavus*	–	x	x	x	–
Carolina madtom	*N. furiosus*	–	x	–	–	–
Orangefin madtom	*N. gilberti*	–	x	x	–	–
Tadpole madtom	*N. gyrinus*	x	x	x	x	x
Margined madtom	*N. insignis*	x	x	x	x	x
Speckled madtom	*N. leptacanthus*	x	–	–	–	–
Broadtail madtom	*Noturus* new species	x	x	–	–	–
Flathead catfish	*Pylodictis olivaris*	x	x	x	–	–
Pikes	Esocidae					
Redfin pickerel	*Esox americanus*	x	x	x	x	x
Northern pike	*E. lucius*	–	x	x	x	–
Muskellunge	*E. masquinongy*	–	x	x	x	x
Chain pickerel	*E. niger*	x	x	x	x	x
Mudminnows	Umbridae					
Eastern mudminnow	*Umbra pygmaea*	x	x	x	x	x
Trouts	Salmonidae					
Rainbow trout	*Oncorhynchus mykiss*	x	x	x	x	x
Sockeye salmon	*O. nerka*	–	x	–	–	–
Brown trout	*Salmo trutta*	x	x	x	x	x
Brook trout	*Salvelinus fontinalis*	x	x	x	x	x

TABLE 4. *Continued*

Common name	Scientific name	S.C.	N.C.	Va.	Md.	Del.
Pirate perches	Aphredoderidae					
Pirate perch	*Aphredoderus sayanus*	x	x	x	x	x
Cavefishes	Amblyopsidae					
Swampfish	*Chologaster cornuta*	x	x	x	–	–
Topminnows and killifishes	Fundulidae					
Northern studfish	*Fundulus catenatus*	–	–	x	–	–
Golden topminnow	*F. chrysotus*	x	–	–	–	–
Banded killifish	*F. diaphanus*	x	x	x	x	x
Lined topminnow	*F. lineolatus*	x	x	x	–	–
Speckled killifish	*F. rathbuni*	–	x	x	–	–
Waccamaw killifish	*F. waccamensis*	–	x	–	–	–
Bluefin killifish	*Lucania goodei*	x	x	–	–	–
Livebearers	Poeciliidae					
Eastern mosquitofish	*Gambusia holbrooki*	x	x	x	x	x
Least killifish	*Heterandria formosa*	x	x	–	–	–
Sailfin molly	*Poecilia latipinna*	x	x	–	–	–
Silversides	Atherinidae					
Brook silverside	*Labidesthes sicculus*	x	–	x	–	–
Inland silverside	*Menidia beryllina*	x	x	x	x	x
Waccamaw silverside	*M. extensa*	–	x	–	–	–
Sticklebacks	Gasterosteidae					
Fourspine stickleback	*Apeltes quadracus*	–	x	x	x	x
Threespine stickleback	*Gasterosteus aculeatus*	–	x	x	x	x
Sculpins	Cottidae					
Black sculpin	*Cottus baileyi*	–	–	x	–	–
Mottled sculpin	*C. bairdi*	x	x	x	x	x
Banded sculpin	*C. carolinae*	–	x	x	–	–
Slimy sculpin	*C. cognatus*	–	–	x	x	–
Potomac sculpin	*C. girardi*	–	–	x	x	–
Broadband sculpin	*Cottus* new species (3)	–	–	x	–	–
Temperate basses	Moronidae					
White perch	*Morone americana*	x	x	x	x	x
White bass	*M. chrysops*	x	x	x	–	–
Striped bass	*M. saxatilis*	x	x	x	x	x
Sunfishes	Centrarchidae					
Mud sunfish	*Acantharchus pomotis*	x	x	x	x	x
Roanoke bass	*Ambloplites cavifrons*	–	x	x	–	–
Rock bass	*A. rupestris*	x	x	x	x	x
Flier	*Centrarchus macropterus*	x	x	x	x	–
Blackbanded sunfish	*Enneacanthus chaetodon*	x	x	x	x	x
Bluespotted sunfish	*E. gloriosus*	x	x	x	x	x
Banded sunfish	*E. obesus*	x	x	x	x	x
Redbreast sunfish	*Lepomis auritus*	x	x	x	x	x

TABLE 4. *Continued*

Common name	Scientific name	S.C.	N.C.	Va.	Md.	Del.
Green sunfish	*L. cyanellus*	x	x	x	x	x
Pumpkinseed	*L. gibbosus*	x	x	x	x	x
Warmouth	*L. gulosus*	x	x	x	x	x
Bluegill	*L. macrochirus*	x	x	x	x	x
Dollar sunfish	*L. marginatus*	x	x	–	–	–
Longear sunfish	*L. megalotis*	–	x	x	x	–
Redear sunfish	*L. microlophus*	x	x	x	–	–
Spotted sunfish	*L. punctatus*	x	x	–	–	–
Redeye bass	*Micropterus coosae*	x	x	–	–	–
Smallmouth bass	*M. dolomieu*	x	x	x	x	x
Spotted bass	*M. punctulatus*	–	x	x	–	–
Largemouth bass	*M. salmoides*	x	x	x	x	x
White crappie	*Pomoxis annularis*	x	x	x	x	x
Black crappie	*P. nigromaculatus*	x	x	x	x	x
Pygmy sunfishes	Elassomatidae					
Carolina pygmy sunfish	*Elassoma boehlkei*	x	x	–	–	–
Everglades pygmy sunfish	*E. evergladei*	x	x	–	–	–
Bluebarred pygmy sunfish	*E. okatie*	x	–	–	–	–
Banded pygmy sunfish	*E. zonatum*	x	x	–	–	–
Perches	Percidae					
Western sand darter	*Ammocrypta clara*	–	–	x	–	–
Sharphead darter	*Etheostoma acuticeps*	–	x	x	–	–
Greenside darter	*E. blennioides*	–	x	x	x	–
Rainbow darter	*E. caeruleum*	–	–	x	–	–
Bluebreast darter	*E. camurum*	–	–	x	–	–
Greenfin darter	*E. chlorobranchium*	–	x	x	–	–
Carolina darter	*E. collis*	x	x	x	–	–
Fantail darter	*E. flabellare*	x	x	x	x	–
Savannah darter	*E. fricksium*	x	–	–	–	–
Swamp darter	*E. fusiforme*	x	x	x	x	x
Christmas darter	*E. hopkinsi*	x	–	–	–	–
Turquoise darter	*E. inscriptum*	x	x	–	–	–
Blueside darter	*E. jessiae*	–	x	x	–	–
Kanawha darter	*E. kanawhae*	–	x	x	–	–
Longfin darter	*E. longimanum*	–	–	x	–	–
Pinewoods darter	*E. mariae*	x	x	–	–	–
Johnny darter	*E. nigrum*	–	x	x	x	–
Tessellated darter	*E. olmstedi*	x	x	x	x	x
Candy darter	*E. osburni*	–	–	x	–	–
Waccamaw darter	*E. perlongum*	–	x	–	–	–
Riverweed darter	*E. podostemone*	–	x	x	–	–
Redline darter	*E. rufilineatum*	–	x	x	–	–
Maryland darter	*E. sellare*	–	–	–	x	–

TABLE 4. *Continued*

Common name	Scientific name	S.C.	N.C.	Va.	Md.	Del.
Sawcheek darter	*E. serrifer*	x	x	x	–	–
Snubnose darter	*E. simoterum*	–	–	x	–	–
Swannanoa darter	*E. swannanoa*	–	x	x	–	–
Seagreen darter	*E. thalassinum*	x	x	–	–	–
Tippecanoe darter	*E. tippecanoe*	–	–	x	–	–
Variegate darter	*E. variatum*	–	–	x	–	–
Glassy darter	*E. vitreum*	–	x	x	x	–
Wounded darter	*E. vulneratum*	–	x	x	–	–
Banded darter	*E. zonale*	x	x	x	x	–
Duskytail darter	*Etheostoma* new species	–	–	x	–	–
Yellow perch	*Perca flavescens*	x	x	x	x	x
Tangerine darter	*Percina aurantiaca*	–	x	x	–	–
Blotchside logperch	*P. burtoni*	–	x	x	–	–
Logperch	*P. caprodes*	–	x	x	x	–
Channel darter	*P. copelandi*	–	–	x	–	–
Piedmont darter	*P. crassa*	x	x	x	–	–
Gilt darter	*P. evides*	–	x	x	–	–
Appalachia darter	*P. gymnocephala*	–	x	x	–	–
Longhead darter	*P. macrocephala*	–	–	x	–	–
Blackbanded darter	*P. nigrofasciata*	x	–	–	–	–
Stripeback darter	*P. notogramma*	–	–	x	x	–
Sharpnose darter	*P. oxyrhynchus*	–	x	x	–	–
Shield darter	*P. peltata*	x	x	x	x	x
Roanoke logperch	*P. rex*	–	–	x	–	–
Roanoke darter	*P. roanoka*	–	x	x	–	–
Dusky darter	*P. sciera*	–	x	x	–	–
Olive darter	*P. squamata*	–	x	–	–	–
Sauger	*Stizostedion canadense*	x	x	x	–	–
Walleye	*S. vitreum*	x	x	x	x	x
Drums	Sciaenidae					
Freshwater drum	*Aplodinotus grunniens*	–	x	x	–	–
Total species		142	210	209	104	80

Distribution in the region is given both in the written accounts and on range maps. Range maps were generated from distributions figured in D. S. Lee et al., *Atlas of North American Freshwater Fishes* (1980); E. F. Menhinick, *The Freshwater Fishes of North Carolina* (1991); and Page and Burr, *A Field Guide to Freshwater Fishes.* Many species are restricted to only a few river drainages, and these drainages are mentioned. However, there has been transfer of fishes from native to other waters through deliberate and accidental introductions, such as when live bait that was purchased or caught in one body of water was released into another at the end of a fishing trip, and widely introduced species are identified. Luckily, such distributional changes are mostly limited to game species and occur primarily in quiet and heavily fished waters, such as reservoirs, lakes, and ponds.

The relative abundance of a species is provided when known. Abundance may vary widely, however, from locality to locality and from year to year.

Habitat information provided will indicate where to look for a given species. This too varies with water quality and other factors.

The natural history section includes information on food, growth, and reproduction. The most common food is indicated if it is known. The life history of the species is outlined, usually with an indication of when and where breeding occurs and with a discussion of selected aspects of breeding when these are known. Temperatures are given in degrees Fahrenheit. Information is included on fishing and utilization of sport or commercially important species.

Preceding each species account, or group of species accounts, is a family account. The latter provides an overview of the larger group to which each of the groups in the region belongs. Because the science of classification of fishes is still an ongoing area of research, the numbers of species mentioned in the family accounts is subject to future changes. This applies also to the data on biology.

Although it might seem that the biology of the fishes of the mid-Atlantic region is well known, such is not the case, especially for the smaller nongame species. Many details need study. There is still great opportunity for the nature-oriented nonprofessional and the professional biologist to make significant observations on size, habitat, food, longevity, nest construction, spawning behavior, movements, and other aspects of life history. The authors will appreciate receiving from readers information on distribution and record sizes or on details of biology not mentioned in this book. They can be contacted at the following addresses:

Rudolf G. Arndt
Faculty of Natural Sciences and Mathematics
The Richard Stockton College of New Jersey
Pomona, NJ 08240

David G. Lindquist and James F. Parnell
Department of Biological Sciences
University of North Carolina at Wilmington
Wilmington, NC 28403

Fred C. Rohde
North Carolina Division of Marine Fisheries
127 Cardinal Drive Extension
Wilmington, NC 28405

This key to the families of freshwater fishes of the mid-Atlantic region applies best to adult or near-adult individuals. The key consists of a sequence of pairs of contrasting statements (couplets), each of which is preceded by a number and letter combination (for example, 1A and 1B). To use this key, read the first couplet and see which of the two choices better applies to your specimen. Then read the number within the key indicated for that choice, proceed to that couplet, and repeat the process until you arrive at the appropriate family. (The drawings below each family will help you make your identification.) Then turn to the plates indicated for photographs of species within that family and the text page indicated for information on the relevant species.

• 1A. Body elongate and snakelike; dorsal fin long **2**
1B. Body not elongate and snakelike; dorsal fin variable **3**
• 2A. Jaws present; pectoral fins present; with one pair of gill openings; up to five feet long **Freshwater eels** (Pl. 11, p. 71)

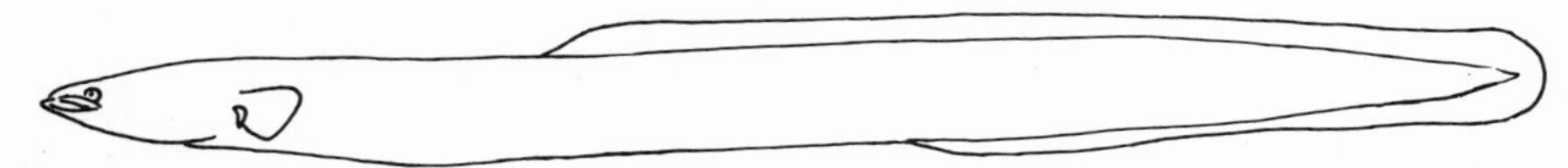

2B. Jaws absent; mouth a disc-shaped funnel; with seven pairs of gill openings; no pectoral fins; up to four feet long **Lampreys** (Pls. 1–4, p. 57)

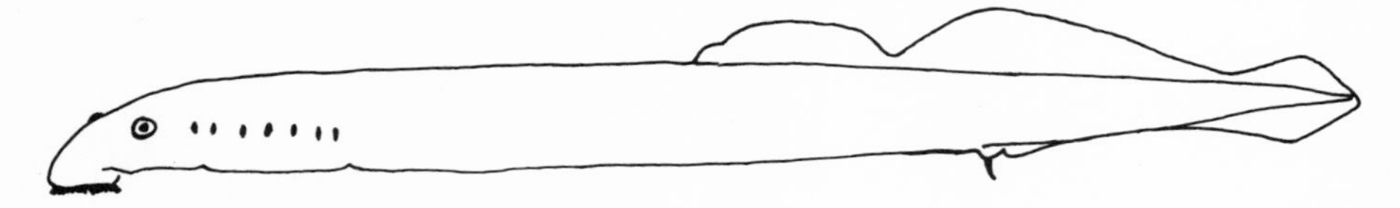

• 3A. Tail fin with upper lobe much longer than lower (sharklike); mouth entirely behind the front of the eye **4**
3B. Tail fin not sharklike; mouth at least partially ahead of the eye **5**
• 4A. Body with five rows of bony plates; snout long, flat, and triangular; up to 14 feet long **Sturgeons** (Pl. 5, p. 61)

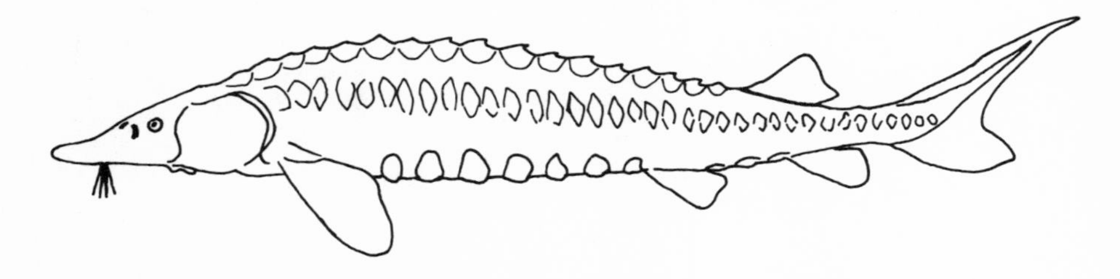

4B. Body scaleless; snout extremely long and paddle-shaped; up to seven feet long **Paddlefishes** (Pl. 6, p. 63)

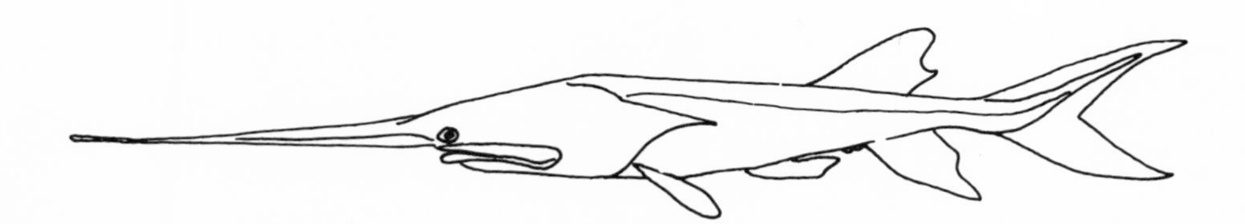

- 5A. Single dorsal fin, spines absent or only one present **6**
  5B. Single dorsal fin with three or more spines, or dorsal fin separated into two parts **19**
- 6A. Dorsal fin long, more than half of the total length; gular plate present; up to 3½ feet long **Bowfins** (Pls. 8–9, p. 67)

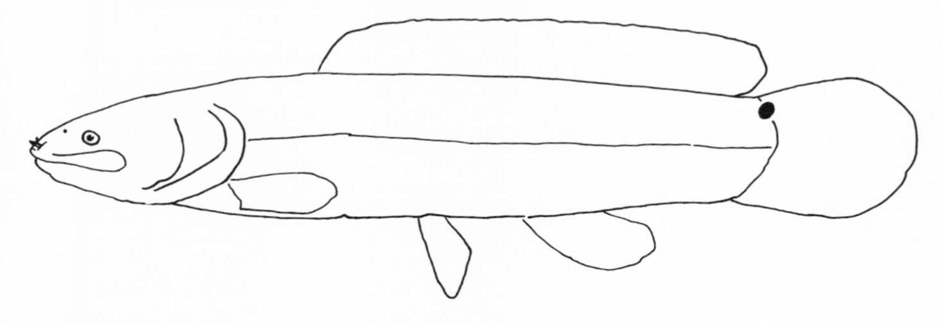

  6B. Dorsal fin short, comprising less than half of the total length **7**
- 7A. Body elongate with armorlike cover of hard scales; snout elongate, with many long, sharp teeth; up to six feet long **Gars** (Pl. 7, p. 65)

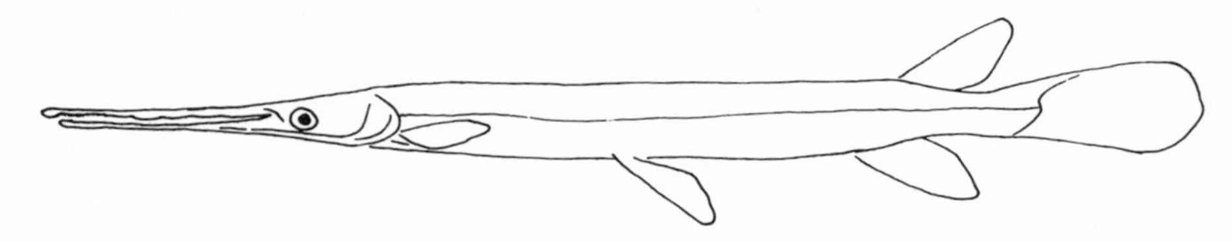

  7B. Body not elongate; no armorlike cover of scales; snout short **8**
- 8A. Anus (in adult) located in throat region, between the gills; tail fin rounded **9**
  8B. Anus located in normal position (before anal fin); tail fin variable **10**
- 9A. Pelvic fins absent; no spines in dorsal or anal fins; up to three inches long **Cavefishes** (Pl. 114, p. 139)

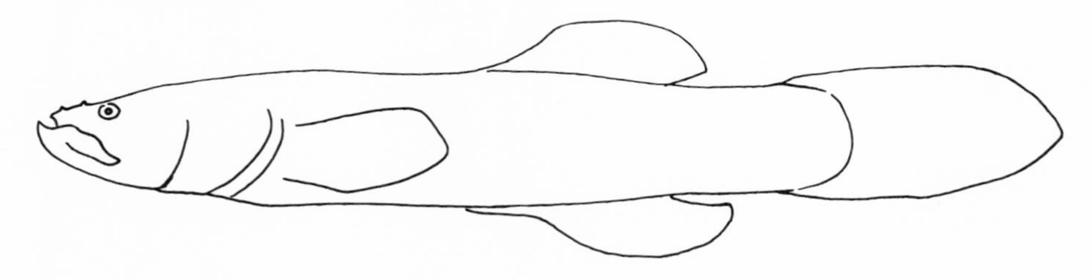

9B. Pelvic fins present; spines in dorsal and anal fins; up to six inches long **Pirate perches** (Pl. 113, p. 137)

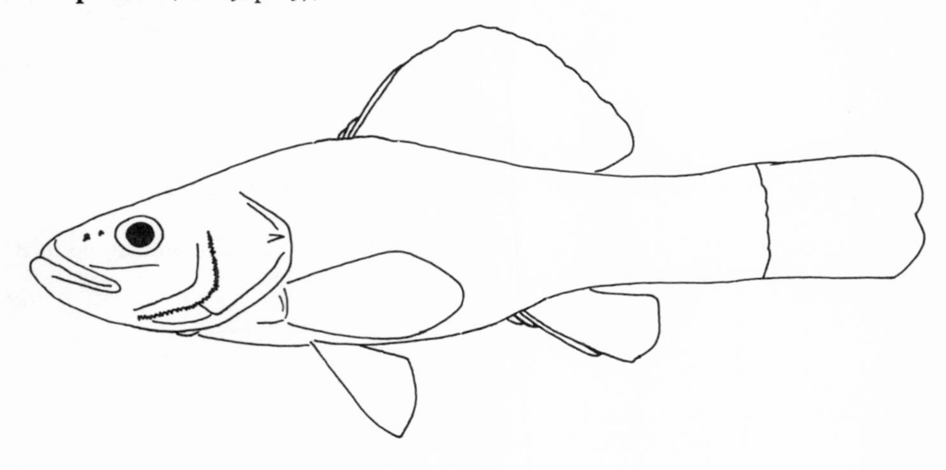

- 10A. Adipose fin present **11**

  10B. Adipose fin absent **12**
- 11A. Scales on body absent; eight large barbels present on head near mouth; spines in dorsal and pectoral fins; up to 5½ feet long **Bullhead catfishes** (Pls. 92–105, p. 119)

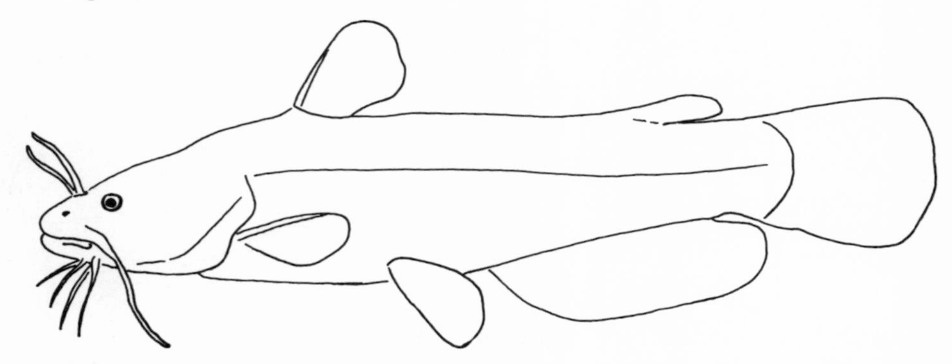

11B. Scales present; no barbels; no spines; up to 2¾ feet long **Trouts** (Pls. 110–12, p. 133)

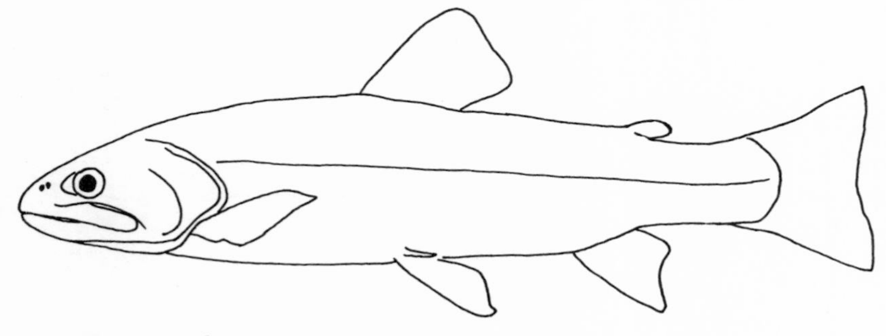

• 12A. Tail fin rounded; lateral line absent; scales present on top of head **13**
12B. Tail fin forked; lateral line variable; scales absent on top of head **15**
• 13A. Mouth terminal; upper jaw not protractile; up to four inches long **Mudminnows** (Pl. 109, p. 131)

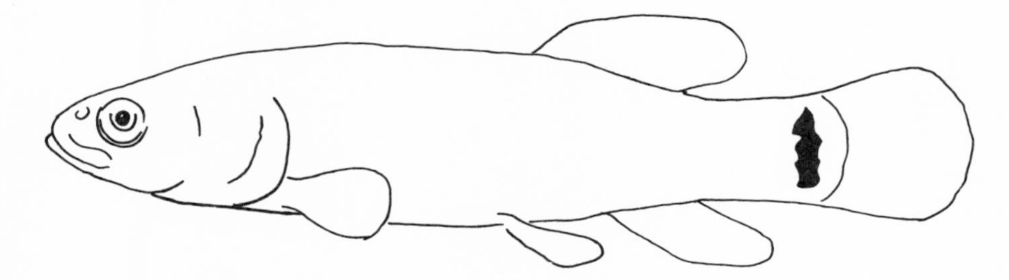

13B. Mouth superior; upper jaw protractile **14**
• 14A. Dorsal fin located over anal fin; anal fin of male rounded; up to seven inches long **Topminnows and killifishes** (Pls. 115–21, p. 140)

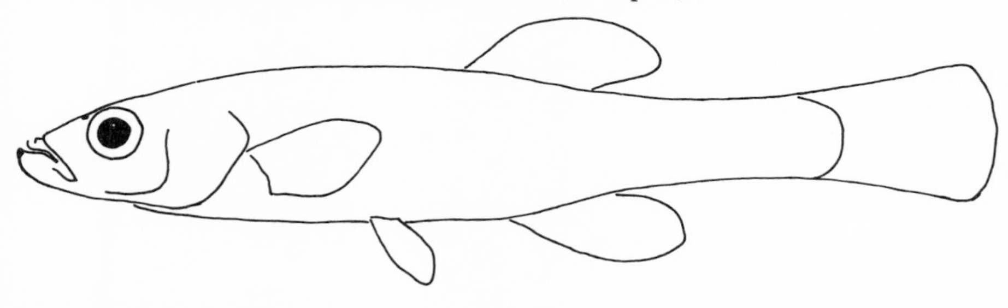

14B. Dorsal fin located behind anal fin; anal fin of male slender and rod-like; up to six inches long **Livebearers** (Pls. 122–24, p. 146)

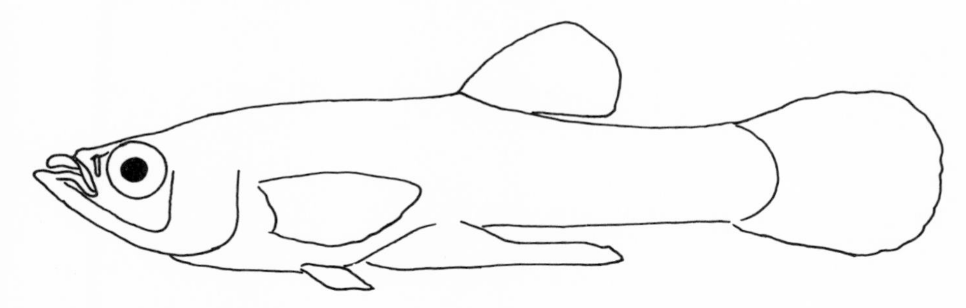

• 15A. Belly with a knifelike keel; adipose eyelid present **16**
15B. Belly rounded (a naked keel present posterior to pelvic fins in golden shiner); adipose eyelid absent **17**
• 16A. Lateral line absent; keel with sharp projecting scales; anal fin located behind dorsal fin; up to 2½ feet long **Herrings** (Pls. 12–13, p. 73)

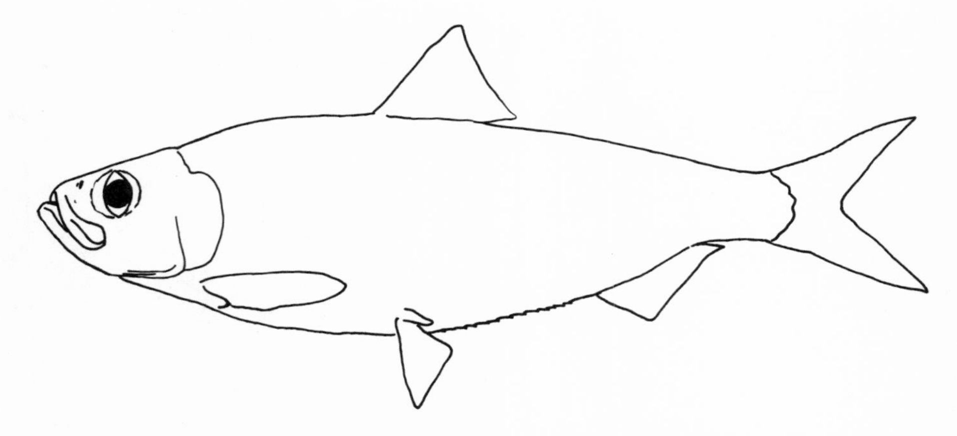

16B. Lateral line present; keel smooth; anal fin located below dorsal fin; up to 1½ feet long **Mooneyes** (Pl. 10, p. 69)

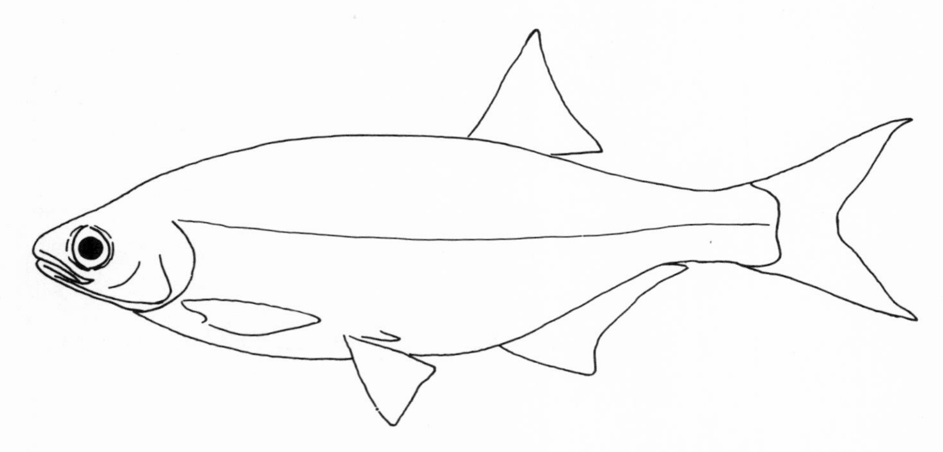

• 17A. Snout elongate and ducklike; scales small; up to six feet long **Pikes** (Pls. 106–8, p. 128)

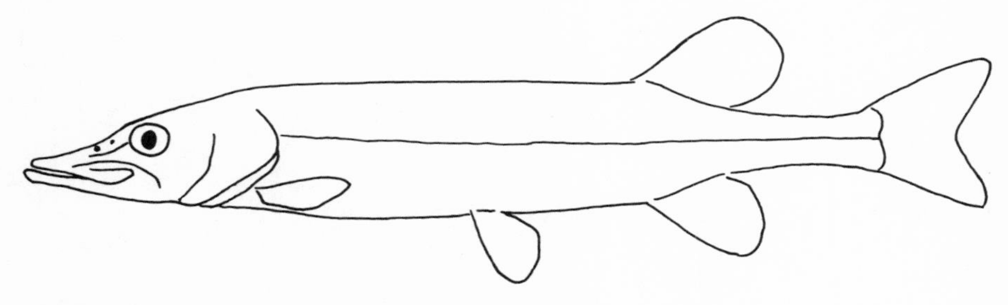

17B. Snout not ducklike; scales large **18**

• 18A. Lips thick and fleshy; dorsal fin with ten or more rays; up to 2½ feet long **Suckers** (Pls. 78–91, p. 107)

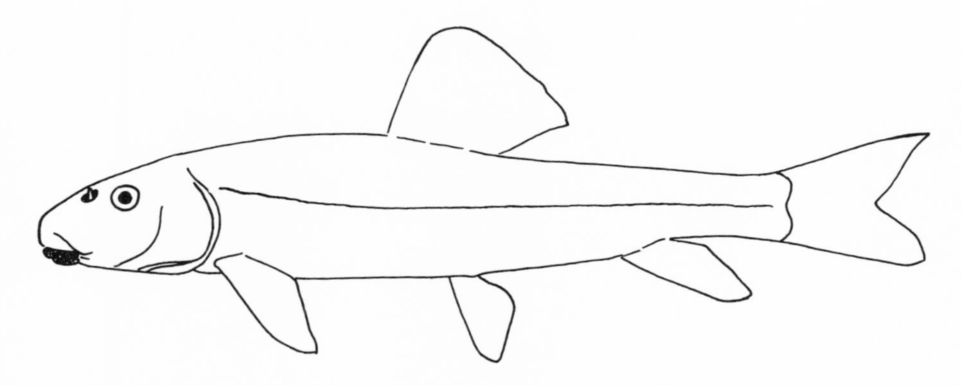

18B. Lips thin and smooth; dorsal fin with eight or nine rays, or if more, a thick spine present at front; up to four feet long **Carps and minnows** (Pls. 14–77, p. 77)

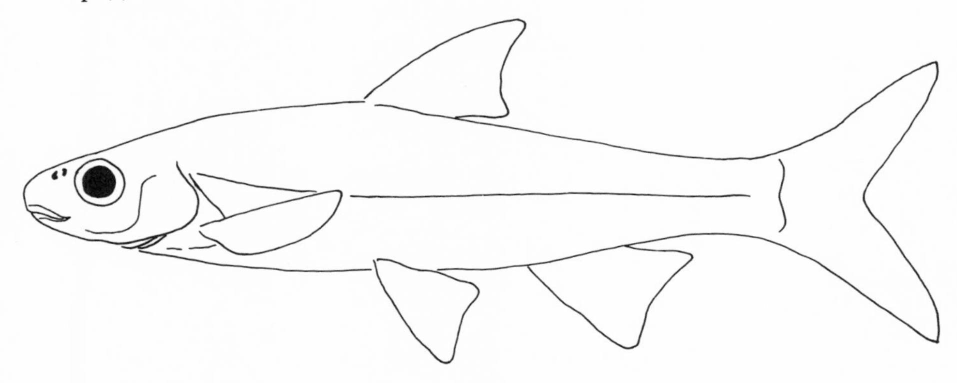

- 19A. Body without scales **20**
  19B. Body with scales **21**
- 20A. Dorsal fin with three to five stout spines separated from each other; up to 2½ inches long **Sticklebacks** (Pl. 127, p. 153)

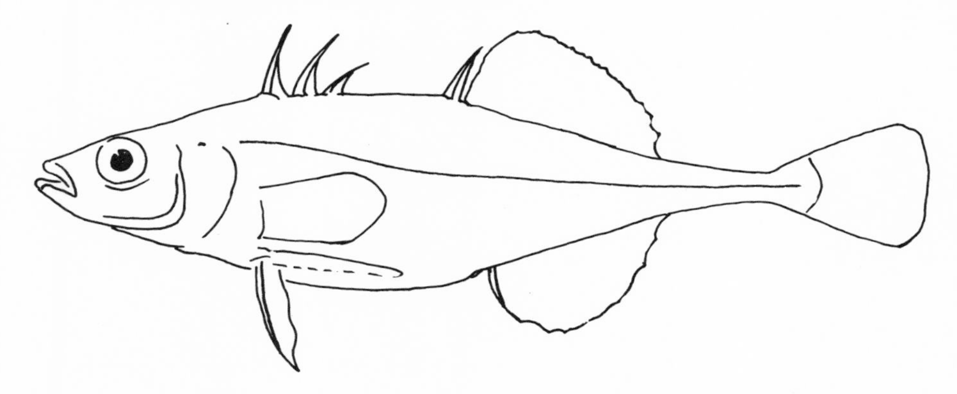

20B. Dorsal fin with weak spines connected by membrane; up to six inches long **Sculpins** (Pls. 128–29, p. 155)

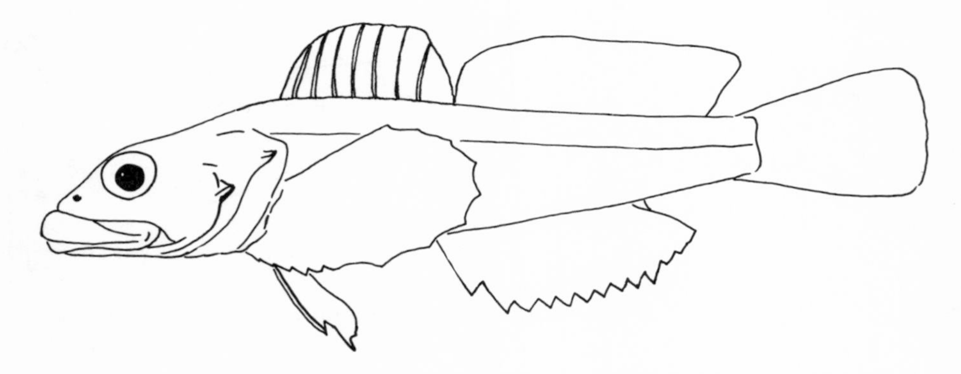

• 21A. Pelvic fins abdominal; dorsal fins widely separated; body elongate, delicate, and silvery; up to five inches long **Silversides** (Pls. 125–26, p. 150)

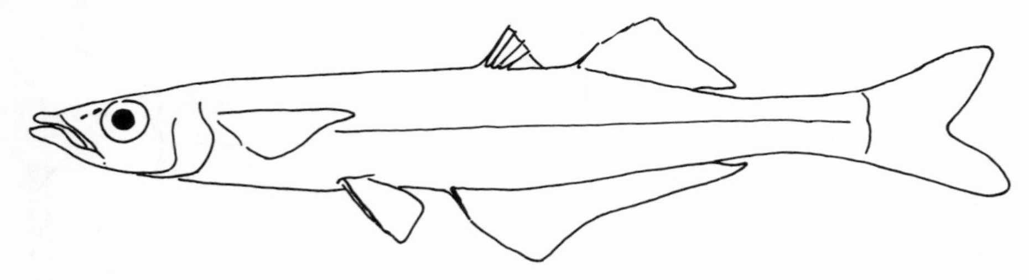

21B. Pelvic fins thoracic; dorsal fins joined or slightly separated **22**

• 22A. Anal fin with one or two spines **23**

22B. Anal fin with three or more spines **24**

• 23A. Lateral line extends onto tail fin; second dorsal fin much longer than first dorsal fin and with more than 25 rays; upto 3¼ feet long **Drums** (Pl. 203, p. 203)

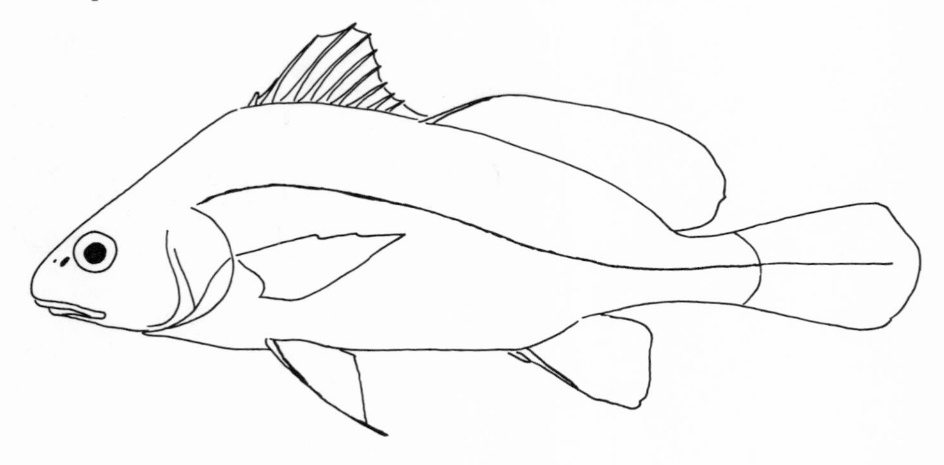

23B. Lateral line absent or present; if latter, not present on caudal fin; second dorsal fin usually not longer than first dorsal fin and with fewer than 25 rays; up to three feet long **Perches** (Pls. 158–202, p. 178)

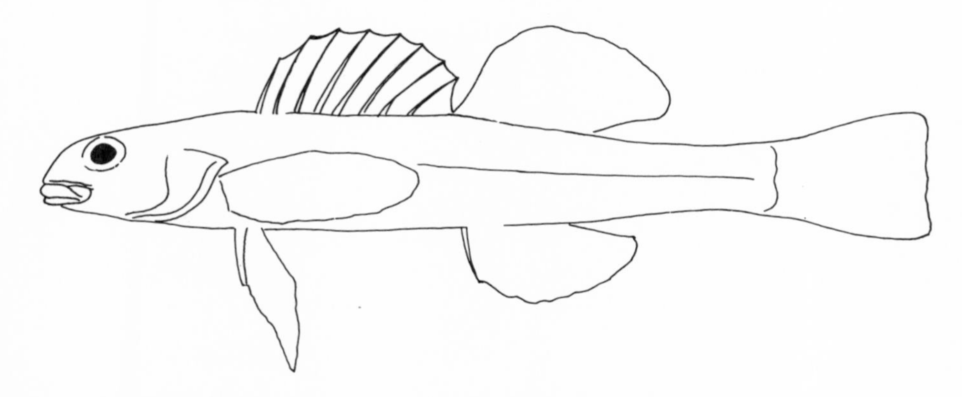

• 24A. Dorsal fins separated or only slightly connected; sharp spine on rear of gill cover; up to 6½ feet long **Temperate basses** (Pls. 130–32, p. 157)

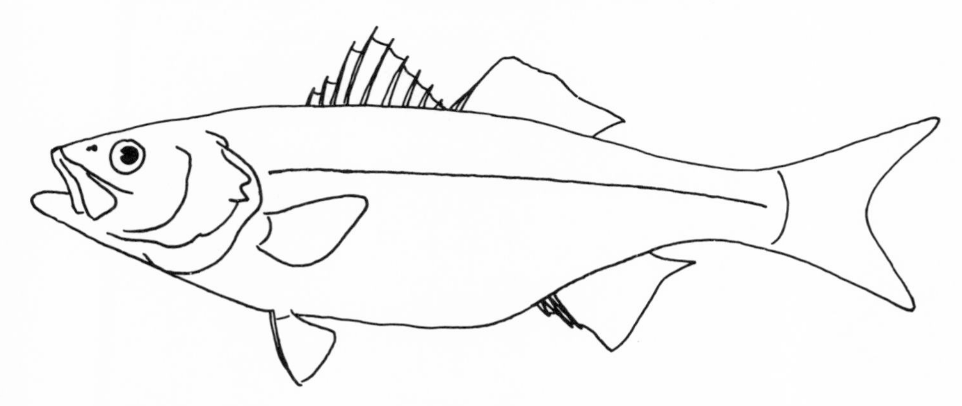

24B. Dorsal fins well connected, and a notch may be present; no spine on rear of gill cover **25**

• 25A. Lateral line present; dorsal fin with 6 to 12 spines; up to three feet long **Sunfishes** (Pls. 133–53, p. 161)

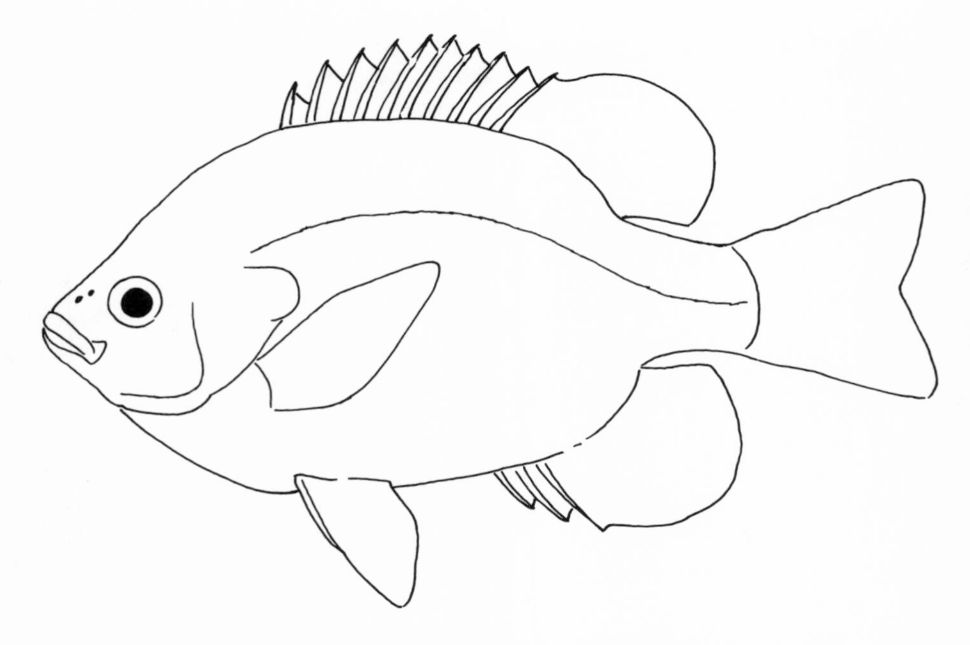

25B. Lateral line absent; dorsal fin with three to five spines; up to 1½ inches long
Pygmy sunfishes (Pls. 154–57, p. 176)

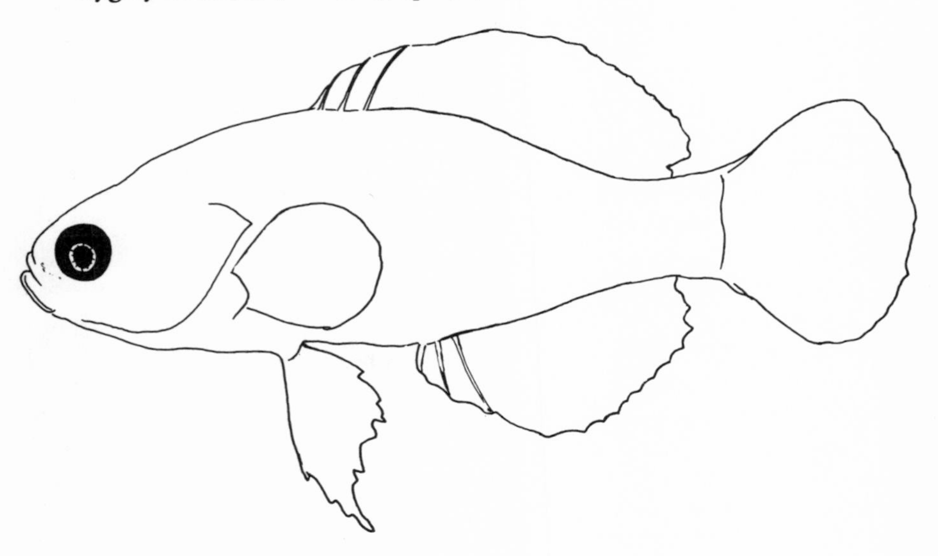

Lampreys *Family Petromyzontidae*

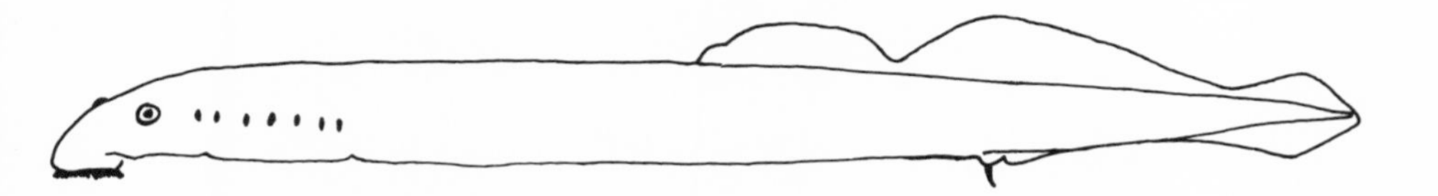

This family includes a total of 41 species, and it is found in eastern and western North America, most of Europe, northern Asia, the temperate parts of South America and Australia, and all of New Zealand. Adult size ranges between 3½ inches and almost four feet. With their slender and snakelike body, with slimy smooth skin, lampreys superficially resemble the common American eel, but they differ from it and all other fishes in the region by their lack of paired fins, scales, jaws, enamaloid teeth, paired nostrils, and bones. Rather, they have only a long and low fin on the back and underside that is connected to the tail, an oval sucking disc with horny, keratinoid teeth (which also occur on the tongue), one nostril located on the midline on the top of the head, seven pairs of clearly visible external gill openings, and a skeleton composed of highly flexible soft cartilage.

Lampreys are fascinating, although their appearance and/or reputation make them repugnant to many people. They are an ancient group, of which fossils 280 million years old and essentially identical to living forms are known. Whether to consider them as fishes or as fishlike vertebrates is a current area of debate among scientists. These animals have been much studied, primarily because of their unusual life histories and because of the well-publicized parasitism of the sea lamprey.

All lampreys begin life in fresh water. Some species then become adult in fresh water, while others migrate into and mature in salt water. Of the former group, some do not feed as juveniles, while some are parasitic on fishes, and all species in the latter group are parasitic on fishes before they return to spawn in fresh water.

The life cycle of all lampreys is similar and consists of three stages: egg, larva, and adult; some species also have a juvenile stage. All reproduce in a particle substrate such as sand/gravel or gravel/rubble, depending on species size, in freshwater riffles. Eggs hatch into larvae, which in the case of lampreys are known as ammocoetes. Ammocoetes of all species in the region occupy sediment-bottomed pools located near their spawning areas. All ammocoetes are semiburrowing filter feeders on plankton, bacteria, and detritus. An overhanging and flexible oral hood partly covers the mouth and helps direct water into it. Ammocoetes may remain as larvae for up to seven years. Transformation or metamorphosis from larva to adult, or juvenile, occurs over a period of several months from late summer to fall. External changes then include development of an oral sucking disc, "teeth," eyes, a genital papilla in the male, larger fins, and adult colors. All lampreys die after they spawn.

Lampreys have an undeserved bad reputation owing to the inadvertent introduction of the sea lamprey into the Great Lakes through the Welland Canal, a series of locks built in the early 1800s to allow ships to bypass Niagara Falls, which resulted in its massive depredation from the 1940s to the 1960s on commercially important fishes such as burbot, rainbow and lake trouts, and whitefishes. Some other species of lampreys are also parasitic. Only fishes are parasitized. Some species are considered to be a food delicacy in several localities, and in a few regions they have been so extensively used as fish bait that their continued existence is threatened.

Nineteen species of lampreys are found in the fresh waters of North America; three of these also occur commonly in brackish and salt water. Five species in three genera inhabit the fresh waters of the mid-Atlantic region, one of which also occurs in the region's coastal salt waters. The species in the region follow one of two paths after transformation. Three species cease to feed and live off stored fats, and these are often referred to as "brook" lampreys. Brook lampreys are small, dull-colored, harmless filter feeders and thus are rarely noticed in their habitat of small streams. Two species, the sea lamprey and the Ohio lamprey, become parasitic after the filter feeding ammocoetes stage. After transformation, juveniles of this second group migrate to a larger body of water and attach themselves to a host fish, rasp a hole in the body wall, and feed on the blood and lymph that oozes out. The host fish does not usually die from the attack, but may die as a result of infection. Most populations of parasitic lampreys have reached an equilibrium with their prey, and the massive depredation by the sea lamprey on its host fishes in the Great Lakes was highly unusual.

Adult lampreys can be identified to species by total body length, differences in tooth patterns on the oral disc, number of muscle segments on the body, whether or not the fins are notched, and body color. Larvae have only poorly developed species-specific characters, primarily based on the number of body muscle segments, and they are thus quite difficult to identify, even for the expert.

Ohio lamprey
Ichthyomyzon bdellium
Mountain brook lamprey
Ichthyomyzon greeleyi
Least brook lamprey
Lampetra aepyptera Pl. 1
American brook lamprey
Lampetra appendix Pl. 2
Sea lamprey
Petromyzon marinus Pls. 3–4

Description. Ohio lamprey: 4.9 to 11.8 inches (124 to 300 mm). A parasitic species, the adult is medium-sized and has black lateral line pores, no notch in the dorsal fin, and sharp and well-developed teeth.

Mountain brook lamprey: 4.3 to 6.8 inches (110 to 172 mm). The nonparasitic adult has a single unnotched dorsal fin, black spots along the lateral line, and numerous sharp, long teeth.

Least brook lamprey: 3.5 to 5.9 inches (90 to 150 mm). The nonparasitic adult develops a mottled silver-black coloration on the back. The dorsal fin is notched, there are no black spots on the lateral line, and the few teeth present are low and blunt. It has less than 62 body segments. This species rarely reaches a length of five inches.

American brook lamprey: 3.9 to 7.1 inches (99 to 180 mm). The adult of this nonparasitic species is similar to the least brook lamprey but is larger, has more and sharper teeth, more body segments (63 or more), and a dark blotch on the tail fin. The dorsal fin is notched.

Sea lamprey: 5.1 to 47.2 inches (130 to 1,200 mm). The adult of this largest lamprey is parasitic and has two dorsal fins separated by a deep notch. The oral disc is as wide or wider than the head and holds many large, strong teeth. The body is strongly mottled with dark and light brown or gray. The spawning adult becomes blue-black.

Distribution and Abundance. The Ohio lamprey is restricted within the region to the upper Tennessee River drainage in southwestern Virginia. It occurs sporadically throughout the rest of the Ohio River basin. The mountain brook lamprey occurs in the region in tributaries of the upper Tennessee River in North Carolina and Virginia. It is widespread in the nearby Ohio River basin. The American brook lamprey occurs throughout much of eastern North America and in all states except South Carolina in the mid-Atlantic region. It occurs in tributaries of the upper Tennessee River and on the Atlantic slope from northern Delaware to southern Virginia. It is considered threatened in North Carolina. The least brook lamprey occurs widely in the eastern United States, and in the mid-Atlantic region is found on the Atlantic slope from northern Delaware to east central North Carolina. It is listed as of special concern in North Carolina. The sea lamprey is found in streams and rivers along the Atlantic coast from Labrador to Florida. Landlocked populations exist in all the Great Lakes and in several lakes in New York. It also occurs in Europe.

Abundance depends on time of the year and location. Adults are often common when they spawn in late winter or early spring, and ammocoetes can occur in the many hundreds in their pool habitat. But because adults are ephemeral and drab-colored, and the ammocoetes occur in muddy pool areas of streams, few people ever see them.

Habitat. The adult of the Ohio lamprey occurs in medium-sized to large rivers. The adult of the mountain brook lamprey occurs in much smaller streams, in riffles with a gravel and rubble bottom, where they excavate nests on the bottom. The least brook lamprey is found

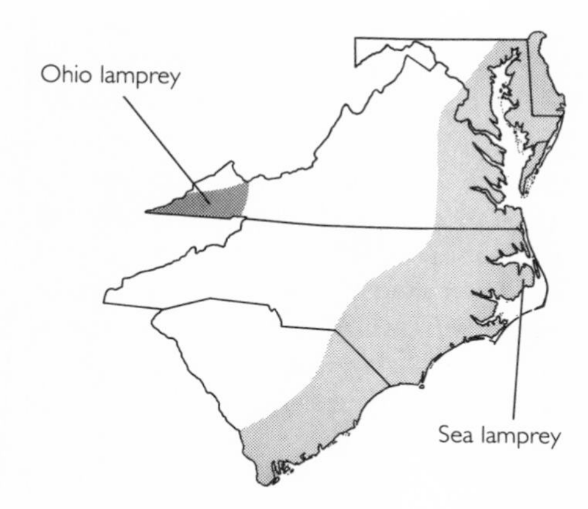

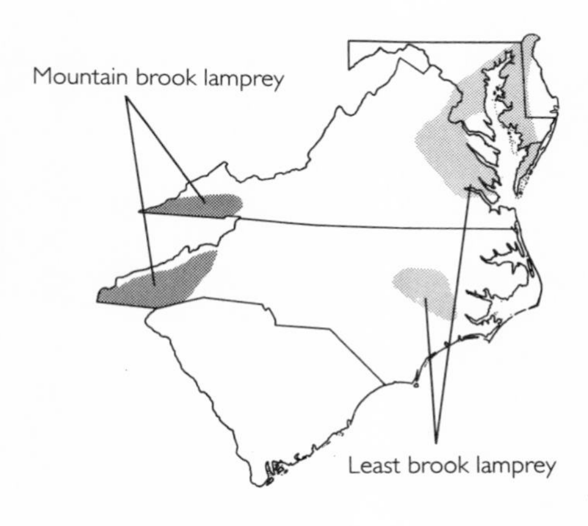

in warm, mostly slow, sandy, slightly acidic, and small creeks. The American brook lamprey in the mountains occurs in streams similar to those of the mountain brook lamprey but along the coast is found in creeks similar to those inhabited by the least brook lamprey. Juveniles of the sea lamprey migrate downstream into estuaries or into the ocean in the mid-Atlantic region. They remain in these coastal areas for up to two years and then return to spawn in riffles in small rivers and large creeks.

Natural History. Life history strategies of brook lampreys in the region are similar. Spawners excavate nests in riffle areas by moving sand and gravel with their sucking disc and by lateral undulations of the body, and many individuals are usually involved. Nests are round excavations, deepest in the center, about as wide as the adults are long, and readily visible against the silt-covered adjacent gray bottom. Individuals of mid-Atlantic species are usually active in groups and construct a number of nests adjacent to one another. The number of eggs deposited per female ranges from many hundreds in the small nonparasitic species to many hundreds of thousands for the largest parasitic species, the sea lamprey.

Metamorphosis begins in late summer and is completed by mid-fall. After transformation, the sea lamprey is parasitic on sharks, Atlantic menhaden, American shad, gizzard shad, trouts, chain pickerel, catfishes, crappies, bluefish, weakfish, and mackerels. The parasitic stage lasts for up to 28 months. Spawning migrations to small rivers and large creeks begin in the last months of life, from March through June. Both sexes participate in the construction of nests in gravel/rock riffles. Spawners are typically monogamous.

The sea lamprey has achieved notoriety because of the collapse of commercial fish stocks after its invasion of the Great Lakes (except Lake Ontario, where it occurred naturally). While the sea lamprey was no doubt a major pest, the large commercial fishery then occurring in these lakes was also an important contributor to the decline of fish stocks. The sea lamprey was once common along the Atlantic coast; in recent decades, however, it has declined because of efforts to eradicate it and because of the depletion of its major host species, and it has not had any serious economic impact in recent years.

Sturgeons *Family Acipenseridae*

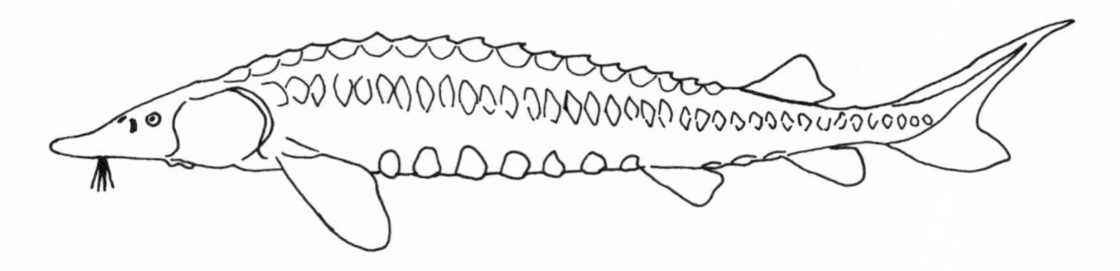

Sturgeons are a small group of 24 species widely distributed in the Northern Hemisphere. The group is at least 200 million years old. These large and primitive fishes have four whiskerlike barbels across the front of a protrusible mouth located under a long, pointed snout. Five rows of large, bony plates (scutes), separated by naked skin, run the length of the body. The upper lobe of the tail extends much farther back than the lower lobe and resembles the asymmetrical (heterocercal) tail of a shark. Some species reach a length of 14 feet or greater. Sturgeons once formed the basis for important fisheries along the East Coast, but overfishing and degradation of rivers have resulted in their decline. Their roe, made into caviar, is a well-known delicacy, and their meat is also tasty and highly valued.

Shortnose sturgeon
Acipenser brevirostrum
Atlantic sturgeon
Acipenser oxyrhynchus Pl. 5

Description. Two species are found along the East Coast of the United States.

Shortnose sturgeon: 16.9 to 42.9 inches (430 to 1,090 mm). This fish has a relatively short snout and wide mouth, and scutes are usually absent between the anal fin and the lateral row of scutes.

Atlantic sturgeon: 34.6 to 169.3 inches (880 to 4,300 mm). Fishermen sometimes refer to this species as the sharpnose sturgeon because it has a pointed snout with a narrow mouth. The snout is relatively blunt, however, in larger specimens. This can cause confusion with the shortnose sturgeon, but the latter attains a maximum length of only 3½ feet. Several scutes are usually present between the anal fin and the lateral row of scutes.

Distribution and Abundance. Both sturgeons are found in rivers and in oceanic waters along the East Coast. The Atlantic sturgeon is more widely distributed. The Atlantic sturgeon once sustained a substantial commercial fishery in the Atlantic coast states, but catches have declined strongly, and most states have prohibited fishing for it in recent years; in North Carolina it is listed as of special concern. The shortnose sturgeon is listed as endangered by the federal government.

Habitat. Sturgeons are bottom dwellers and prefer deep waters and a soft substrate. At spawning time they require freshwater areas with a fast flow and a rough bottom.

Natural History. Sturgeons are anadromous and spend much of their life in estuaries and the ocean and move into freshwater rivers to spawn. Both species spawn from February to May. The shortnose sturgeon lays some 30,000 eggs, while the Atlantic sturgeon lays up to several million. The shortnose sturgeon matures when three years old, while the Atlantic sturgeon does not mature until age seven. The young of the latter may remain in fresh water for up to five years before they migrate to the ocean. The Atlantic sturgeon reaches an age of over 50 years, while southern populations of the shortnose sturgeon rarely attain 30 years. Both sturgeons commonly eat worms, crustaceans, insect larvae, small clams, and small fishes.

Shortnose sturgeon

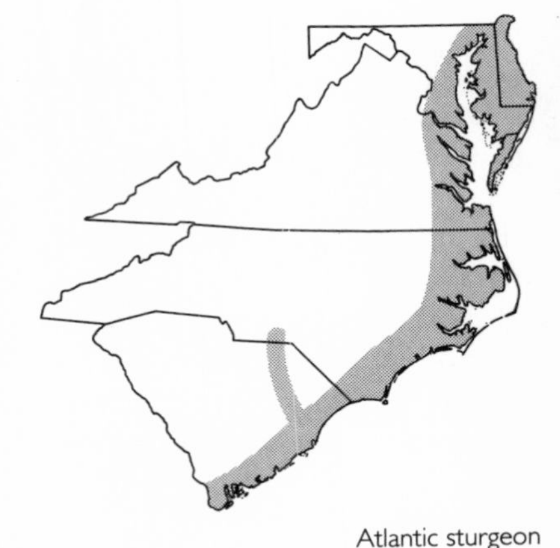

Atlantic sturgeon

Paddlefishes *Family Polyodontidae*

Paddlefishes are primitive and characterized by a unique large projecting paddle-shaped snout. They have several sharklike characteristics, such as a heterocercal tail, a large underslung mouth, and a specialized sharklike intestine with a spiral valve; but they are not related to sharks and are inoffensive to man. They feed on zooplankton which they strain out of the water. There are only two similar-appearing species, one found in the American heartland and the other in the Yangtze River in China.

Paddlefish

Polyodon spathula Pl. 6

Description. 35 to 87 inches (890 to 2,210 mm). The paddlefish is large, brown, and sharklike, with a snout that comprises about one-third of the total length. It has a large mouth on the underside of the head and a greatly elongated, pointed, fleshy gill cover. It can reach a weight of 198 pounds and a length of over six feet. It looks like no other fish in the mid-Atlantic area.

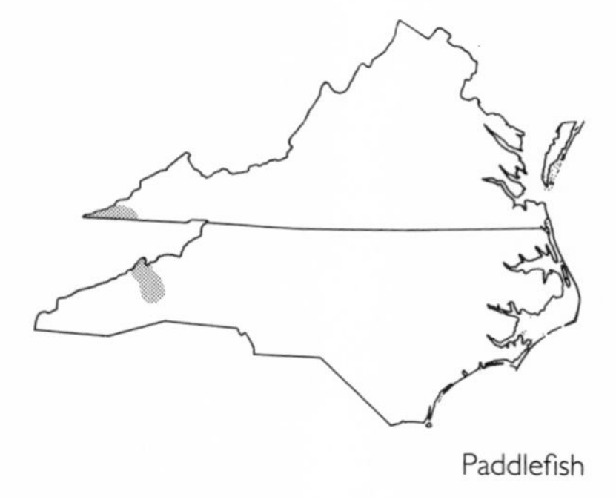

Paddlefish

Distribution and Abundance. In the region it is known only from reports in the 1800s from the French Broad River near Asheville, North Carolina, and from recent reports from the Clinch and Powell rivers in extreme western Virginia. It is widely distributed in the Mississippi River and its larger tributaries, from the Great Lakes to the upper Missouri River and south to Texas and Alabama. Its distribution has been greatly reduced, primarily due to river damming. However, it is still locally common.

Habitat. It formerly occurred primarily in free-flowing large rivers and adjacent flooded bayous. Today it inhabits mainly large and altered rivers and impoundments.

Natural History. The paddlefish feeds as it swims with its huge mouth gaped open and strains zooplankton out of the water with its numerous enlarged, close-set, and sievelike gill rakers. It spawns from March to June, after a move from calmer to flowing waters, especially over a gravel bottom. Thousands of eggs are released, sink, and adhere to the gravel. The hatchling looks like the closely related sturgeons but quickly develops the paddlelike rostrum. Growth is rapid, and sexual maturity is first attained when eight years old, when the male has reached a weight of at least 15 pounds and the female 30 pounds. An age of 30 years has been reported. The paddlefish is often caught by snagging it. Its flesh is edible, and the roe is tasty and much in demand, and consequently high-priced. The paddlefish is considered to be endangered in North Carolina and threatened in Virginia.

Gars *Family Lepisosteidae*

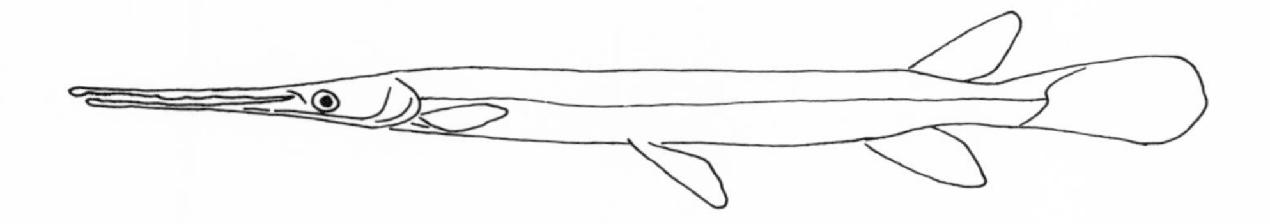

This small group of primitive fishes is found only in the New World, from southeastern Canada to Cuba and Central America, and it is primarily a freshwater group. There are two living genera, *Lepisosteus* and *Atractosteus.* Of a total of seven species in the group, five occur in North America, but only one of these is found in the mid-Atlantic region. Gars have a long and slender body covered by a thick protective sheath of diamond-shaped, nonoverlapping, hard ganoid scales, and long jaws with numerous needlelike teeth. They have many primitive features such as a lunglike air bladder and a spiral valve in the intestine. Adults are predators, almost exclusively on other fishes. The alligator gar, *A. spatula*, found west of the region, reaches a length of about 11 feet and a weight of almost 300 pounds.

Longnose gar

Lepisosteus osseus Pl. 7

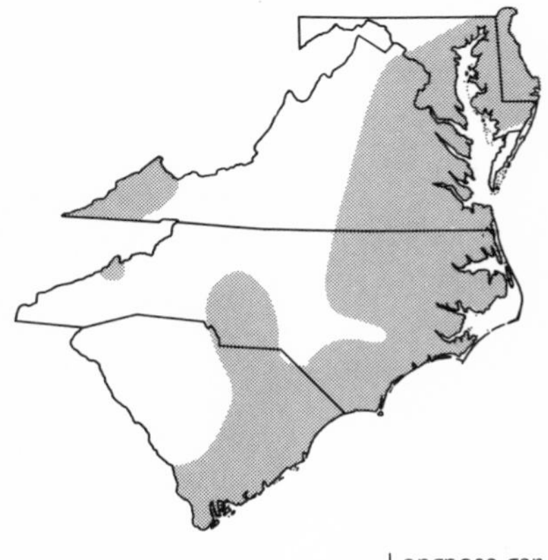

Longnose gar

Description. 19.7 to 72.0 inches (500 to 1,830 mm). The longnose gar has a long and slender snout, which is more than twice as long as the rest of the head. The back and side are olive brown to green, with scattered dark oval spots on the fins and body, and the venter is yellowish-white. The young gar has a broad dark stripe on the side and a more narrow one on the back. The tail of very young specimens ends in a long, fleshy filament.

Distribution and Abundance. The longnose gar is generally common throughout its range, which includes the Atlantic and Gulf coastal plains south into northeastern Mexico as well as the Great Lakes and Mississippi River. In the mid-Atlantic region it occurs throughout the coastal plain, where it is common and sometimes occurs in brackish water, and in tributaries of the Tennessee River.

Habitat. It occurs in backwaters, pools, lakes, and rivers, where it is frequently associated with weedy areas. It often swims in open water just under the surface.

Natural History. The longnose gar is noted for its ability to rise to the surface and gulp air to fill an air bladder, which then serves as an accessory respiratory organ, or a primitive "lung." It feeds mostly at night and near the surface. A primary food is the gizzard shad. Adept at stealthy stalking, the longnose gar moves slowly alongside an unsuspecting fish, like a stick drifting with the current, and then whips its jaws sideways and grasps the prey with its long needle-like teeth. The prey is then maneuvered through a series of flips and thrusts headfirst into the mouth and swallowed.

Adults congregate in spring to spawn in shallow weedy or rocky areas. Each female may be pursued by as many as 15 males and will spawn with two to six males simultaneously at irregular intervals. A female ready to spawn will lead the males in a large loop that corresponds to the size of the spawning area. A male then nudges the underside of the female with his snout, and the spawning group stops and orients with snouts close to the bottom. Rapid violent quivering accompanies the actual release of eggs and sperm, and thrashing and splashing at the surface is then common. The eggs are large, approximately one-eighth to one-sixth inch in diameter, greenish, and attach to the substrate. A large female can produce more than 36,000 eggs in one season. They are abandoned, but are toxic and thus are apparently protected from predators. The eggs hatch in three to nine days into fry with a large yolk sac, and they attach to submerged objects with an adhesive organ on the snout. After about nine days the yolk sac is absorbed, and the young becomes a predator that can take prey nearly one-third its length. The male matures in three to four years and the female in about six years. The male may attain an age of over 20 years and the female an age of more than 30 years. The longnose gar helps to control rough and game fishes.

The longnose gar is considered to be a nongame species and a nuisance bait stealer by hook-and-line fishermen, and it often creates terrible tangles when it wanders into the nets of gill netters. It does, however, provide sport for the bow fisherman, and it is sometimes caught in a snare made of piano wire or unraveled rope with a minnow attached. The meat is white, boneless, and can be smoked with delicious results. It has gained popularity and has been successfully marketed in Arkansas.

The young longnose gar is a fascinating aquarium fish, although it requires a little more space than other species its size. It will accept a variety of live food.

Bowfins *Family Amiidae*

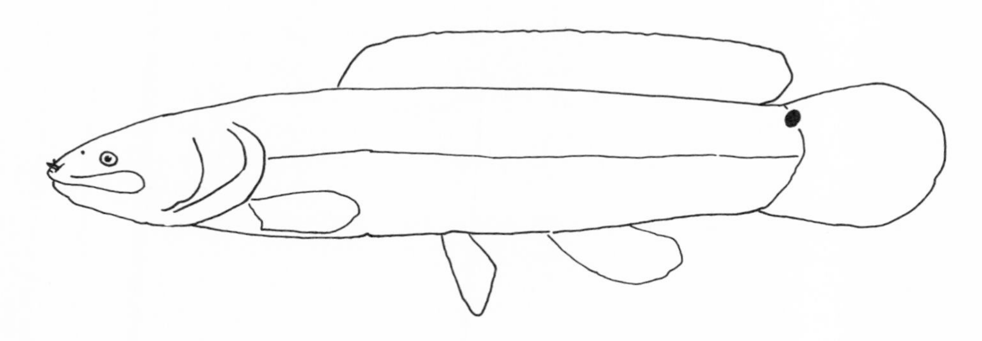

The bowfin is the only living member of this family, and it is today restricted to eastern North America. There are many extinct forms known from Europe, Greenland, Africa, and South America, which lived over 100 million years ago. While some extinct forms were marine, the genus *Amia* has apparently always lived only in fresh water.

Bowfin

Amia calva Pls. 8–9

Description. 18.0 to 42.9 inches (457 to 1,090 mm). The body is moderately long, of about the same height from head to tail, and approximately round in cross section. The dorsal fin is long and low; the caudal fin is rounded. The back and side are olive, with dark mottled patches. There are numerous sharp teeth. A large black spot on the base of the caudal fin is prominent in the young, and it is ringed with orange or yellow in the mature male. A unique large, bony plate, the gular plate, is located on the underside of the lower jaw. The species reaches a weight of almost 20 pounds.

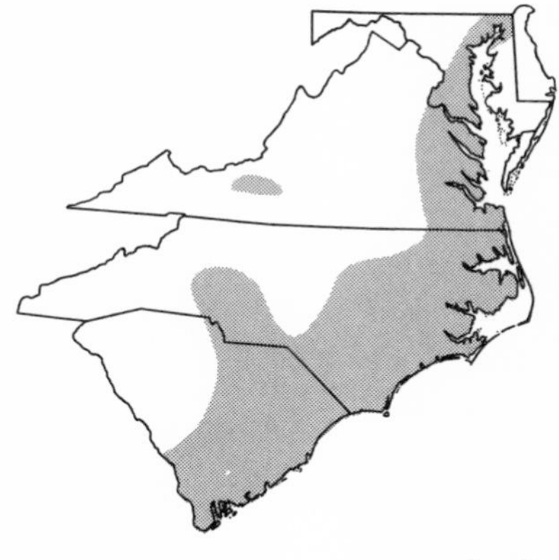

Bowfin

Distribution and Abundance. The bowfin is known by a dozen mostly disparaging names, which include blackfish, dogfish, grinnel, and mudfish. It is found throughout most of the eastern half of the United States and in southeastern Canada. It is generally common throughout much of the coastal plain in the mid-Atlantic region but is absent from Delaware and the Eastern Shore of Maryland and Virginia. The bowfin has been introduced into piedmont and mountain reservoirs in Virginia and North Carolina.

Habitat. It prefers dense vegetation and clear water in a variety of swampy habitats such as ditches, channels, borrow pits, pools, and sluggish creeks and rivers.

Natural History. The bowfin spawns in spring when the male begins an elaborate ritual and moves into shallow, weedy waters to prepare a nest. He clears a circular area one to two feet wide by biting and tearing out vegetation and makes a trough four to eight inches deep among roots. He defends this nest and surrounding area against intruding males. A female is attracted into the nest. She lies there as the male circles her for 10 to 15 minutes and occasionally nips her snout or side. He then moves beside her and both strongly beat their fins as they release eggs and sperm into the trough. Four or five batches of adhesive eggs are deposited over a period of about an hour and a half during bouts of alternate circling and spawning. The male guards the nest and fans the eggs with his pectoral fins. He may spawn with several females, and the nest may contain from 2,000 to 5,000 eggs in various stages of development. The eggs begin to hatch in eight to ten days. The fry attach to vegetation for seven to nine days with an adhesive organ on the snout. They swim and feed when one-half inch long, following, and forming a ball of young around, the male. The male guards this mass of fry for several weeks until they are about four inches long.

The bowfin is usually considered a pest when caught by fishermen, although some seek it as a scrappy fighter. It is sometimes eaten and is occasionally served in restaurants under a delicious-sounding pseudonym.

A small bowfin makes an interesting and colorful aquarium fish. It is tolerant of a variety of conditions and readily eats earthworms and commercially prepared foods.

Mooneyes *Family Hiodontidae*

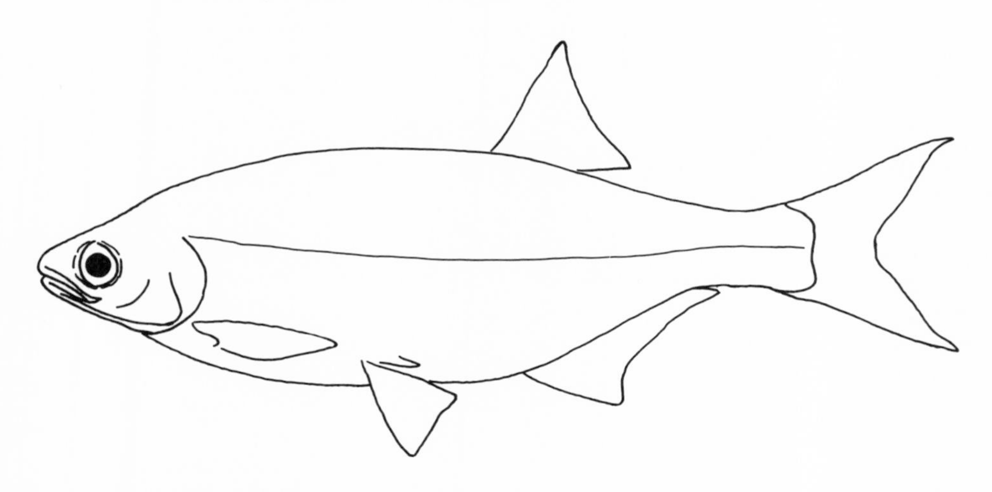

This family contains two similar species, the mooneye and the goldeye, *Hiodon alosoides.* It is restricted to central Canada and the central United States. While both the mooneye and the goldeye superficially resemble the American shad, these silvery, strongly compressed, soft-rayed fishes possess a lateral line and lack a toothed keel on the belly. They live in large creeks, rivers, lakes, and impoundments and feed on insects, crustaceans, mollusks, fishes, and even mice. The female reportedly can produce many thousands of semibuoyant eggs. Sexual maturity is probably reached in three years. Both species were in the past caught with nets and eaten extensively by man. Although not generally sought, they provide good sport to the angler.

Mooneye

Hiodon tergisus Pl. 10

Mooneye

Description. 9.4 to 18.5 inches (240 to 470 mm). Similar in appearance to the shad, the mooneye is differentiated from it by a relatively much larger eye, a dorsal fin much further back on the body (its origin is well behind that of the pelvic fin rather than in advance of it as in the shad), a complete lateral line (which is absent in the shad), and a lack of a toothed and sharp keel on the belly (which is present in the shad).

Distribution and Abundance. In the mid-Atlantic region the mooneye now is known from only the lower 15 or so miles of the French Broad River in western North Carolina, but it ranged further upstream in earlier years. It is widely distributed in rivers in the Mississippi River basin, in adjacent basins to the north, and in Canada. While locally common today, it is reported as being formerly abundant. It is a species of special concern in North Carolina.

Habitat. This fish prefers larger rivers with clear water, usually in fast deep water and over a firm bottom. It also occurs in lakes and impoundments.

Natural History. The mooneye apparently feeds mostly near the surface in swift water. In North Carolina it probably spawns from February to early May. A female deposits some 10,000 to 20,000 eggs, which are about one-sixteenth inch in diameter. Maturity is attained in three to four years, although some individuals can live for up to eight years. Food is mostly insects, although crustaceans, mollusks, and fishes are also taken by the mooneye, which feeds primarily by sight. It was once abundant over much of its range, but it is today much less common as a result of increased siltation, turbidity, and other types of pollution.

Freshwater Eels *Family Anguillidae*

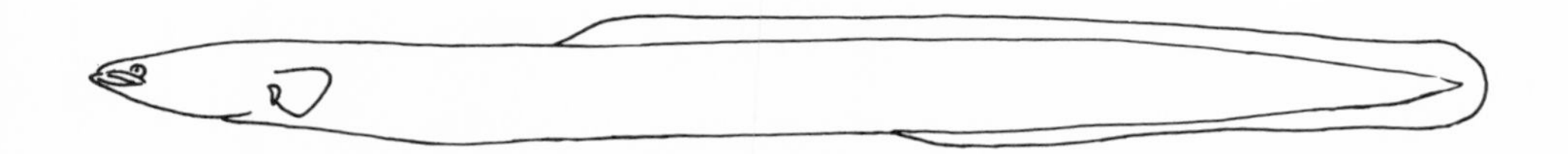

The common and widespread American eel is the best known of 16 species in this family, all of which are in the genus *Anguilla.* The family occurs widely in tropical and temperate seas with the exception of the south Atlantic and the eastern Pacific oceans. Most species begin their life in the ocean, then migrate into fresh water, where they spend most of their life, and finally return to the ocean, where they spawn and then die. This type of spawning migration is known as catadromy. The freshwater eels are the only group of catadromous fishes in North America.

American eel

Anguilla rostrata Pl. 11

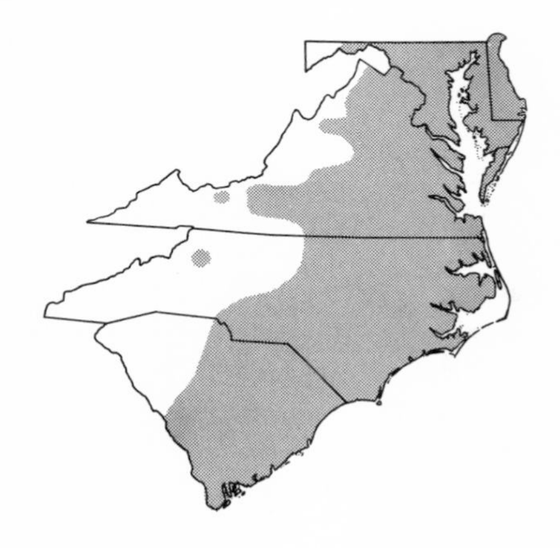

American eel

Description. 24.0 to 59.8 inches (610 to 1,520 mm). This medium-sized to large fish has a long, snakelike body that is gray to brown in color. The dorsal, caudal, and anal fins are united and form a contiguous, low fin that frames the posterior two-thirds of the body; the caudal fin is bluntly rounded. Pelvic fins and a pelvic girdle are absent, the pectoral fin is located high on the side, and the gill slit is small and located just in front of the pectoral fin. The body appears naked, but small cycloid scales are buried in the skin. This and a body that is covered by mucus make it very slippery and have given rise to the expression "slippery as an eel." When caught in a net, it attempts to move toward water with purposeful snakelike movements and does not flip around like other fishes.

Distribution and Abundance. This fish is widespread throughout the mid-Atlantic region and is common in inshore waters, estuaries, rivers, creeks, lakes, and ponds. It occurs in all of the Atlantic Ocean drainages of Canada from Newfoundland south, throughout the eastern half of the United States, in the Caribbean Sea to Colombia, and on Caribbean islands.

Habitat. The American eel prefers areas with a soft bottom such as mud or mud/sand and with vegetation or other shelter in which it can hide. Because of its wanderings and migrations, however, it can be encountered almost anywhere.

Natural History. The American eel begins life as it hatches from eggs laid by parents in deep ocean waters in the Sargasso Sea, which is located south of Bermuda and northeast of the Bahama Islands. The larvae are carried north by the Gulf Stream to coastal areas of the mid-Atlantic region, where they arrive some five to seven months after spawning. The early larva does not look like the adult, in that it has a strongly laterally compressed body, is transparent, and has a relatively large mouth and teeth. When it has reached the coast, and when it is about two inches long, it has become eel-shaped, with dark eyes and a still-transparent body; it is then called a "glass" eel. Slightly later, when several inches long, it is gray-green and looks like a small adult; it is then known as an elver. The young female has a particularly powerful urge to migrate upstream against a current and swims up rivers and their tributaries, and even migrates across land and over dams, usually at night and during a rain, when conditions minimize the danger of desiccation, to reach lakes and ponds. The habitat of the male is coastal brackish waters.

The eel is primarily a predator and feeds on anything of animal origin, primarily fishes but also insects, crayfishes, snakes, earthworms, and larval lampreys. The eel also scavenges, and it is often caught in pots by fishermen who use bait of all kinds, especially fish and chicken. Eels captured by crabbers are often immediately converted into crab bait for crab pots.

At an age of five to seven years, and when only about two feet long, the male is adult and migrates from coastal waters to the Sargasso Sea. He is joined by the much larger female, which reaches a length of five feet and a weight of 16 pounds, which moves downstream from inland fresh waters. Migrations start in fall, and eels spawn in mid-winter in the Sargasso Sea at a depth of over 1¼ miles. Millions of tiny eggs are produced per female. The adults then die.

The eel is considered a delicacy by many, especially peoples in Asia and Europe. The glass eel, elver, and adult are eagerly sought in this region and shipped to Asia, while smoked adult eel is offered in the larger cities of the United States. The eel is usually caught in baited, fine-mesh pots, which work on the same principle as lobster and crab pots. In order to catch the female as it migrated downriver, Native Americans in the eastern United States partially dammed rivers with stones and then netted or speared the eels as they were forced through the restricted openings.

Herrings *Family Clupeidae*

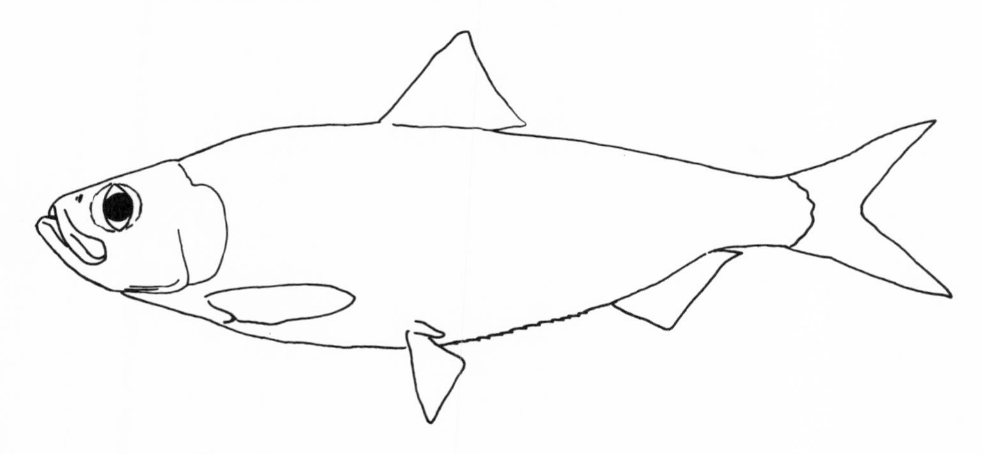

Herrings are easily recognized, silvery, flat-sided fishes, which include sardines. All possess a row of sharp-edged scales on the belly, which creates a sawtooth keel. The eye is partially covered by a transparent membrane, the adipose eyelid. A few species are freshwater, but most are marine, some of which regularly enter fresh water. The last group are anadromous fishes, and they spend most of their adult life in the ocean but return to their natal freshwater streams to spawn. Clupeids occur worldwide, and 6 of the nearly 200 species occur in the fresh waters of the mid-Atlantic region. Several are important commercially, and all are important forage fishes.

Blueback herring
Alosa aestivalis
Hickory shad
Alosa mediocris
Alewife
Alosa pseudoharengus
American shad
Alosa sapidissima Pl. 12

Each spring brings excitement to coastal rivers and streams of the region, as four anadromous fishes—the alewife, blueback herring, American shad, and hickory shad—return to their natal fresh waters to spawn. Both commercial and recreational fishermen then descend upon these waters to harvest these plentiful fishes. Sport fishermen target the two shads, particularly the American shad, and attempt to hook them with small artificial lures. The American shad is also sought by commercial fishermen who use drift gill nets. The alewife and blueback herring, often referred to as river herrings, face gill nets, seines, dip nets, fyke nets, and trawls as they wend their way upstream. The roe of all these species is highly prized, as is the flesh, which is prepared baked, smoked, fried, salted, and pickled.

Description. Blueback herring: 13.8 to 15.7 inches (350 to 400 mm). This river herring can be identified by its bluish back, silvery head, and black peritoneum (the lining of the inside body wall, which can be seen only by making an incision into the body cavity).

Hickory shad: 12.0 to 24.0 inches (305 to 610 mm). This is a medium-sized clupeid and can be identified by its prominently projecting lower jaw and pale (silvery or white) peritoneum. A row of dark spots is present behind the opercle.

Alewife: 9.8 to 15.0 inches (250 to 380 mm). Similar to the blueback herring, the alewife has a gray-blue back, brassy head, and pale (gray) peritoneum (determining the color of the peritoneum is usually critical for an accurate identification).

American shad: 12.0 to 29.9 inches (305 to 760 mm). This is the largest member of the herring family in the United States; the adult averages three to four pounds, but it can weigh eight. Unlike the hickory shad, its lower jaw does not project prominently.

Blueback herring

Distribution and Abundance. All four species are widely distributed along the East Coast and in all river systems in the mid-Atlantic region. The alewife has been introduced into numerous landlocked reservoirs and lakes as a forage fish. However, none of the four species is as abundant as it once was, and the decline has been attributed to overfishing in the ocean, deterioration of water quality, stream channelization, algal blooms, and dams. South Carolina lists the blueback herring and hickory shad as of special concern. The Atlantic coast states currently participate in interstate management actions to help the stocks to recover.

Hickory shad

Habitat. The adults of all species are usually found in the ocean in a narrow band close to their natal streams. Most overwinter in deeper ocean waters offshore, although the American shad makes extensive migrations to and from the Bay of Fundy in Canada in the summer and fall. The spawning areas range from large rivers to creeks only a few inches deep and are reached after a long migration.

Alewife

American shad

Natural History. Most individuals return to their home streams to spawn, for the first time usually when three years old. The spawning peaks occur from the end of March to early May. Spawning is usually accomplished with much splashing as ova and milt are released. Each female lays up to several hundred thousand eggs (up to 600,000 in the American shad), and these hatch in several days. The adults move back to the ocean rapidly after they spawn. The young remain in nursery areas in lower portions of rivers and estuaries until water temperatures decrease in the autumn, when they move out to sea. The young as well as the adults feed on copepods, insects, shrimp, worms, and some fishes. Most growth occurs in the ocean. The maximum age reached is about eight years.

Gizzard shad

Dorosoma cepedianum Pl. 13

Description. 8.8 to 20.5 inches (225 to 520 mm). The gizzard shad, named for its gizzardlike stomach, differs from most other clupeids by the elongated and filamentous last ray of the dorsal fin. It also has a distinctly blunt snout with a subterminal mouth. A similar species, the **threadfin shad**, *D. petenense,* has been widely introduced in the mid-Atlantic region as a forage fish. It can be differentiated from the gizzard shad by a more pointed snout and terminal mouth.

Distribution and Abundance. The gizzard shad occurs throughout almost all of the mid-Atlantic region, as well as in most of the eastern half of the United States, especially in reservoirs, large rivers, and creeks. It forms large and active schools and can become so abundant in reservoirs that it is a nuisance.

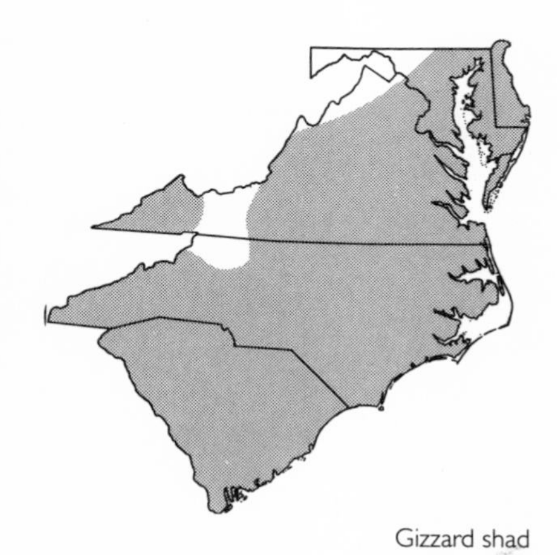
Gizzard shad

Habitat. The gizzard shad inhabits a wide range of waters, such as rivers, creeks, and tailraces below dams, but it prefers calm, warm waters with a high phytoplankton production, such as reservoirs. It is occasionally found in estuaries and the ocean.

Natural History. The gizzard shad can often be seen near the surface, and it frequently leaps or skips out of the water. It strains out its food of algae and other microscopic organisms abundant at the water surface with its long, numerous, and close-set gill rakers on the gill arches. It spawns in spring and early summer;

the eggs are scattered, usually in shallow water, and sink to the bottom, where they adhere to anything they contact. They hatch within four days. Growth of the larvae can be rapid and depends on the fertility of the water. Young gizzard shad are an important food for game fishes, but the young often outgrow predators and then compete with other fishes for space. The gizzard shad is susceptible to rapid environmental changes, such as sudden water temperature changes and depletion of dissolved oxygen, which sometimes result in large die-offs. Few live longer than seven years.

Carps and Minnows *Family Cyprinidae*

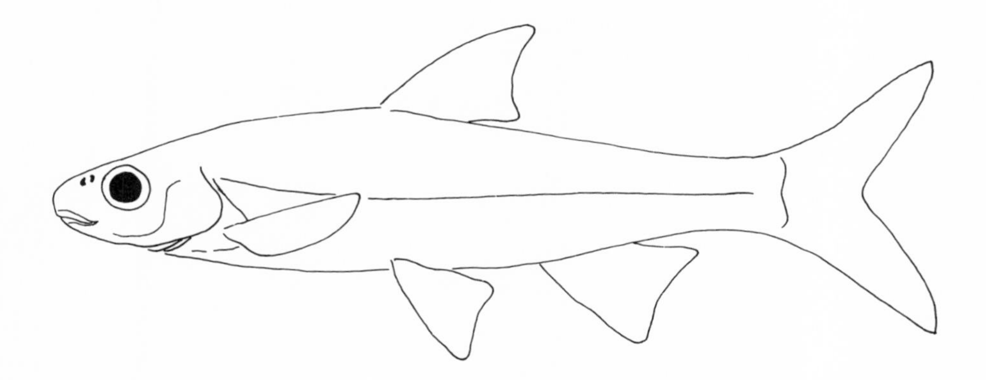

To the average person, any small fish is a minnow, whereas to the ichthyologist only fishes in the family Cyprinidae can properly be referred to by this name. Minnows, as defined by the ichthyologist, and contrary to popular misconception, include many large species, and some in Asia reach a length of almost 10 feet. This is the world's largest family of fishes, containing about 194 genera and almost 2,100 species. The family contains fishes with an extreme range of shapes, sizes, colors, behaviors, and habitats and is thus not easily defined. Members have one dorsal fin, abdominal pelvic fins, cycloid scales, and usually a lateral line. They lack teeth in the mouth, but they do have teeth on bones deep in the throat (pharyngeal teeth), which grind food against a tough pad on to the underside of the skull. The shape and number of these teeth are important to help identify the species, but unfortunately, given the location of the teeth, they can be observed only in dead fishes, and then only after careful dissection. Minnows are strictly freshwater fishes, and they are found over most of North America, Europe, Asia, and Africa. They are important predators, forage fishes, and food for man. Over 10 percent of the total minnow species in the world occur in North America, and 86 species occur in the mid-Atlantic region.

Central stoneroller

Campostoma anomalum Pl. 14

Description. 4.8 to 9.4 inches (122 to 239 mm). This is a round-bodied minnow with a unique distinct and hard cartilaginous ridge present along the edge of the lower jaw. The mouth is located under the head. The intestine is long and looped around the air bladder. The body color is tan to brown. The breeding male becomes orange, especially in the fins, with black bands in the dorsal and anal fins, and the lips become white, especially the upper lip, which also develops a backward-projecting extension and appears as a distinct white moustache; the male then also develops tubercles on most of the head and upper body.

Distribution and Abundance. Widely distributed throughout the central and eastern United States, the central stoneroller is restricted to the mountains and upper piedmont in the mid-Atlantic region. It is common to abundant in high-quality habitat, and some streams in the

Great Smoky Mountains support up to 1,120 fish per acre.

Habitat. This fish is found in rocky riffles, runs, and pools of streams with clear, cool water and a moderate current. It prefers riffles and the runs immediately below them.

Natural History. The central stoneroller scrapes algae off stones with its hard lower lip. Breakdown of this hard-to-digest food is facilitated by its long intestine. The male digs a circular pit nest in mid- to late spring, usually in shallow water at the head of a riffle. He moves pebbles with his mouth or loosens gravel by body movements that allow the current to carry it downstream; large stones are rolled and pushed with the back and head, hence the common name. The eggs are adhesive and lodge among the substrate, and they are not cared for by the parents. The central stoneroller competes with trout for spawning areas, and its digging may sometimes destroy trout redds. But trouts generally spawn earlier than the stoneroller, so competition for space is minimal. The maximum age reached is five years. It is a prized food fish in the Great Smoky Mountains.

Goldfish

Carassius auratus Pl. 15

Description. 4.7 to 16.1 inches (120 to 410 mm). The goldfish in the wild is a cryptic olive color. It has large scales, a long low dorsal fin, and a large and strong spine located at the front of the dorsal and anal fins, each of which follows two much smaller spines. The body is relatively high and carplike but lacks the pair of barbels on each side of the mouth of the carp. In captivity the goldfish is the poodle of the fish world, and it has been selectively bred by the Chinese, Japanese, and Koreans for hundreds if not thousands of years, and more recently by fanciers in other countries as well. This has resulted in the familiar gold color of aquarium goldfish, as well as in black, white, and mottled colors and in an extreme range of body shapes, fin shapes, scales sizes, and eye sizes and locations. Specimens released from bait buckets and those released, or escaped, from aquariums for a while show the colors and/or shapes of their captive heritage, but succeeding generations quickly revert to the drab camouflaging colors and the shape characteristic of the wild, or native, type.

Central stoneroller

Distribution and Abundance. This fish is native to Asia and adjacent Europe. It was long ago brought to North America, and as a result of escapes and well-meaning but ill-advised releases it now occurs at scattered sites in all of the lower 48 states and the adjacent parts of Canada and Mexico. It can be locally common. There are scattered records of it from throughout the mid-Atlantic region; some no doubt represent established populations, while others probably were first-generation releases that survived to be recorded by an observant biologist but then did not successfully reproduce. After extensive sampling in the region, the authors have never encountered it.

Habitat. The goldfish prefers clear, quiet or slow, warm waters such as ponds and the edges of lakes or slow rivers, usually over a bottom of silt and detritus near aquatic vegetation. It can tolerate turbid water, and it often thrives in degraded waters where native species have suffered greatly.

Natural History. The goldfish spawns from late spring through summer when the water temperature reaches 60° and continues as long as the temperature remains above 60° and the population is not overcrowded. When crowded, this fish produces a secretion that represses further spawning. At spawning time several amorous males usually chase a female, and spawning continues for most of the day. The adherent eggs are attached singly—though sometimes in twos or threes—to aquatic plants and other fixed objects and then immediately fertilized. A given fish is reproductive over many weeks or months, and a large female produces several thousand eggs in a season. The time to sexual maturity depends on the strain; some mature in nine months, others not for three or four years. The usual longevity is six to seven years; the maximum age observed is 30 years. It feeds on a wide range of live animals and plants and is also a scavenger. In parks it takes bread and popcorn, and in aquariums it is usually fed prepared dried foods. It can produce viable hybrids with the common carp.

Goldfish

Rosyside dace

Clinostomus funduloides Pl. 16

Description. 2.2 to 4.3 inches (56 to 109 mm). The rosyside dace is separated from others by a large, oblique mouth and a pointed snout. The body is compressed and covered with small scales, and the olive body has a diffuse dark stripe on the side. The breeding male develops a bright red lower side, hence the name "rosyside." A red slash is usually present behind the gill cover in the female and the nonbreeding male. The closely related **redside dace**, *C. elongatus,* has a more pointed snout and a more slender body.

Distribution and Abundance. The rosyside dace occurs in upland drainages on the Atlantic slope from the Delaware River south to the Savannah River and west into the Ohio River basin, including the Tennessee River drainage. It is usually common throughout its range, especially in its preferred habitat. A subspecies of the rosyside dace that occurs in the Little Tennessee River in western North Carolina is considered to be of special concern. The only record of the redside dace within the mid-Atlantic re-

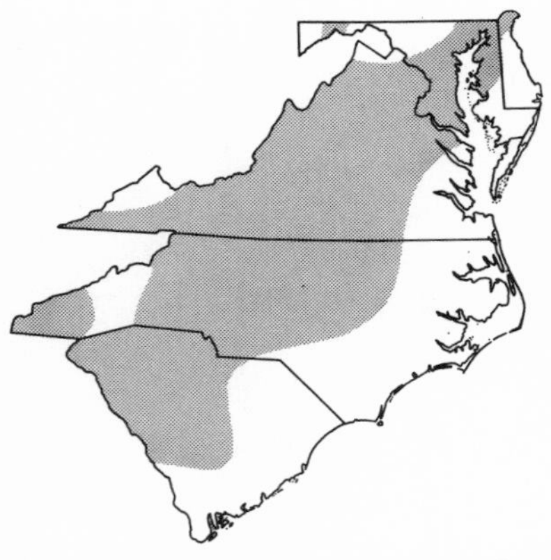
Rosyside dace

gion is from extreme northwestern Maryland.

Habitat. The rosyside dace is most common in flowing pools in clear headwater creeks. It also occurs in large creeks and small rivers.

Natural History. It spawns from early April to late June. A female may lay between 140 and 800 eggs, which are less than one-sixteenth inch in diameter. It grows rapidly and reaches maturity by the end of the first or second year, when about two inches long; the maximum age attained is four years. The diet consists predominantly of insects, both aquatic and terrestrial.

Grass carp

Ctenopharyngodon idella Pl. 17

Description. 29.5 to 49.2 inches (750 to 1,250 mm). The grass carp, also known as the white amur, is an elongate and silvery minnow with large, dark-edged scales and thin lips. The dorsal fin is short and is located at midbody, above the pelvic fins; the paired and anal fins are also short and attached low on the body. Barbels are absent. It attains a weight of 70 pounds in its native waters.

Distribution and Abundance. The grass carp is native to eastern Asia, especially China. The name white amur comes from its occurrence in the Amur River, which forms much of the border between Russia and China. It was introduced into the United States in the 1960s. It now occurs in at least 34 states in the United States, and it is found sporadically throughout the mid-Atlantic region. It is uncommon.

Habitat. This species tolerates a wide range of environmental conditions, including water temperatures of 32° to 95°, salinities of up to ten parts per thousand of total salts, and dissolved oxygen concentrations almost as low as zero parts per million. It is most common in slow, heavily vegetated, and warm water with a muddy bottom, such as ponds, lakes, and slow rivers. In China it also occurs in rice paddies.

Natural History. In China the grass carp spawns from April until mid-August. A female produces hundreds of thousands of eggs, which float. This species feeds extensively on aquatic vegetation, and it can eat its own weight of vegetation in a day; but it no doubt also feeds on small

Grass carp

animal life, and it may scavenge as well. It has been widely introduced in an effort to control rampant vegetation in eutrophic waters and also to serve as a forage species. To reduce the chances of its becoming established as a pest, a nonreproductive strain of it is primarily stocked today. However, the older strain has managed to invade some waters and threatens to become a widespread pest. Fish of the nonreproductive strain that were 10 to 11 inches long when released into a pond in Delaware in April had reached a length of nearly two feet and a weight of five pounds by November. It is often active at the water surface, and it has been reported taken by the osprey.

Satinfin shiner
Cyprinella analostana Pl. 18
Greenfin shiner
Cyprinella chloristia
Whitetail shiner
Cyprinella galactura Pl. 19
Bannerfin shiner
Cyprinella leedsi
Red shiner
Cyprinella lutrensis
Spotfin chub
Cyprinella monacha Pl. 20
Whitefin shiner
Cyprinella nivea Pl. 21
Fieryblack shiner
Cyprinella pyrrhomelas Pl. 22
Spotfin shiner
Cyprinella spiloptera
Steelcolor shiner
Cyprinella whipplei

Description. Ten members of the genus *Cyprinella* occur in the mid-Atlantic region, and all can be distinguished from other shiners by vertical diamond-shaped scales and a black blotch in the dorsal fin. These species are usually difficult to identify, especially as juveniles and females; most readily identified are the large males.

Satinfin shiner: 1.8 to 4.3 inches (47 to 110 mm). The dorsal fin is black speckled on all membranes, and there is a distinct black blotch on the rear half. The breeding male is gray-blue with yellow fins. There are nine rays in the anal fin.

Greenfin shiner: 1.7 to 2.8 inches (44 to 72 mm). Similar to the satinfin shiner, this species has eight instead of nine anal fin rays and a more distinct stripe on the side.

Whitetail shiner: 1.7 to 5.9 inches (44 to 150 mm). This shiner is readily separated from the others by two large creamy white spots located at the base of the caudal fin. The dorsal fin in the spawning male often has a red tint.

Bannerfin shiner: 1.8 to 3.9 inches (46 to 100 mm). The mouth of this shiner is subterminal, and the dorsal fin has a small black blotch at the front; this fin in the breeding male is greatly enlarged and black.

Red shiner: 1.7 to 3.5 inches (44 to 90 mm). This is a deep-bodied species and is further characterized by a blue triangular bar on the side behind the head. The breeding male has a blue body and red fins.

Spotfin chub: 2.6 to 4.2 inches (66 to 108 mm). This species has a slender body that is flattened below, a long snout which extends over the mouth, and a small barbel at the corner of the mouth. The breeding male has two large white bars on the blue side and white edges on the blue fins.

Whitefin shiner: 1.7 to 3.3 inches (44 to 85 mm). Similar to the satinfin shiner, it can be differentiated by the presence of eight anal fin rays, a more slender body, and a dark blue stripe along the side. The male has white fins.

Fieryblack shiner: 1.8 to 4.4 inches

(47 to 110 mm). Distinctively colored, the fieryblack shiner has a red snout, a black bar on the side behind the head, and black edging on the tail fin. The spawning male develops white and red bands on the tail fin.

Spotfin shiner: 1.7 to 4.7 inches (44 to 120 mm). Similar to the satinfin shiner, this species has eight anal fin rays and fewer black specks on the anterior membranes of the dorsal fin.

Steelcolor shiner: 1.8 to 6.3 inches (47 to 160 mm). This shiner can be differentiated from the greenfin shiner by its nine anal fin rays and from the satinfin shiner by a larger number (15 versus 13 or 14) of rays in the pectoral fin. The breeding male has a blue back and side, red snout, yellow fins, and an enlarged dorsal fin.

Distribution and Abundance. The satinfin shiner is found throughout most of the region on the Atlantic slope from the Delaware River to the upper Pee Dee River in South Carolina, as well as north of the region. The greenfin shiner occurs only in the Santee River drainage in southwestern North Carolina and north central South Carolina. The whitetail shiner in the region occurs in the mountains of the Carolinas and southwestern Virginia, where it is common; it also occurs throughout the Tennessee River drainage and in the Ozark Mountains. The bannerfin shiner is found in the coastal plain from the Edisto River in South Carolina and southwest to western Florida. The red shiner is widely distributed in the central United States, but in the mid-Atlantic region it occurs only at several localities in central North Carolina where it has been accidentally introduced. The spotfin chub is restricted to a few tributary systems of the Tennessee River drainage in Virginia, North Carolina, and Tennessee. The whitefin shiner is found only in the coastal plain and piedmont of the Carolinas from the Neuse River in North Carolina to the Savannah River in South Carolina, while the fieryblack shiner occurs in both the Carolinas in the Santee and Pee Dee river drainages almost entirely above the fall line. The spotfin shiner is widely distributed throughout the north central United States and in the mid-Atlantic region ranges from Delaware to western North Carolina. The steelcolor shiner in the region is found only in tributaries of the Tennessee River in extreme southwestern Virginia; it also occurs widely in the Mississippi River basin.

All species are sometimes common and occasionally even abundant in appropriate creeks and rivers, except for the bannerfin shiner, which is uncommon, and the spotfin chub, which is rare and listed as threatened by the federal government.

Habitat. The greenfin, red, whitefin, and spotfin shiners prefer sand and gravel runs and riffles in small to medium-sized rivers, and the spotfin chub, satinfin, whitetail, fieryblack, and steelcolor shiners occur in rocky runs and pools in creeks and small to medium-sized rivers. The bannerfin shiner is found in sandy runs of medium-sized to large rivers.

Natural History. Breeding adult males produce sounds during courtship that may help individuals to recognize members of their own species. The males guard territories but do not care for the nest or young. The females deposit eggs in crevices in submerged objects, such as rocks and logs, a behavior not exhibited by other shiners. They do not release all of their eggs at once, but rather spawn several times during the season, which is in May and June. These shiners are sight

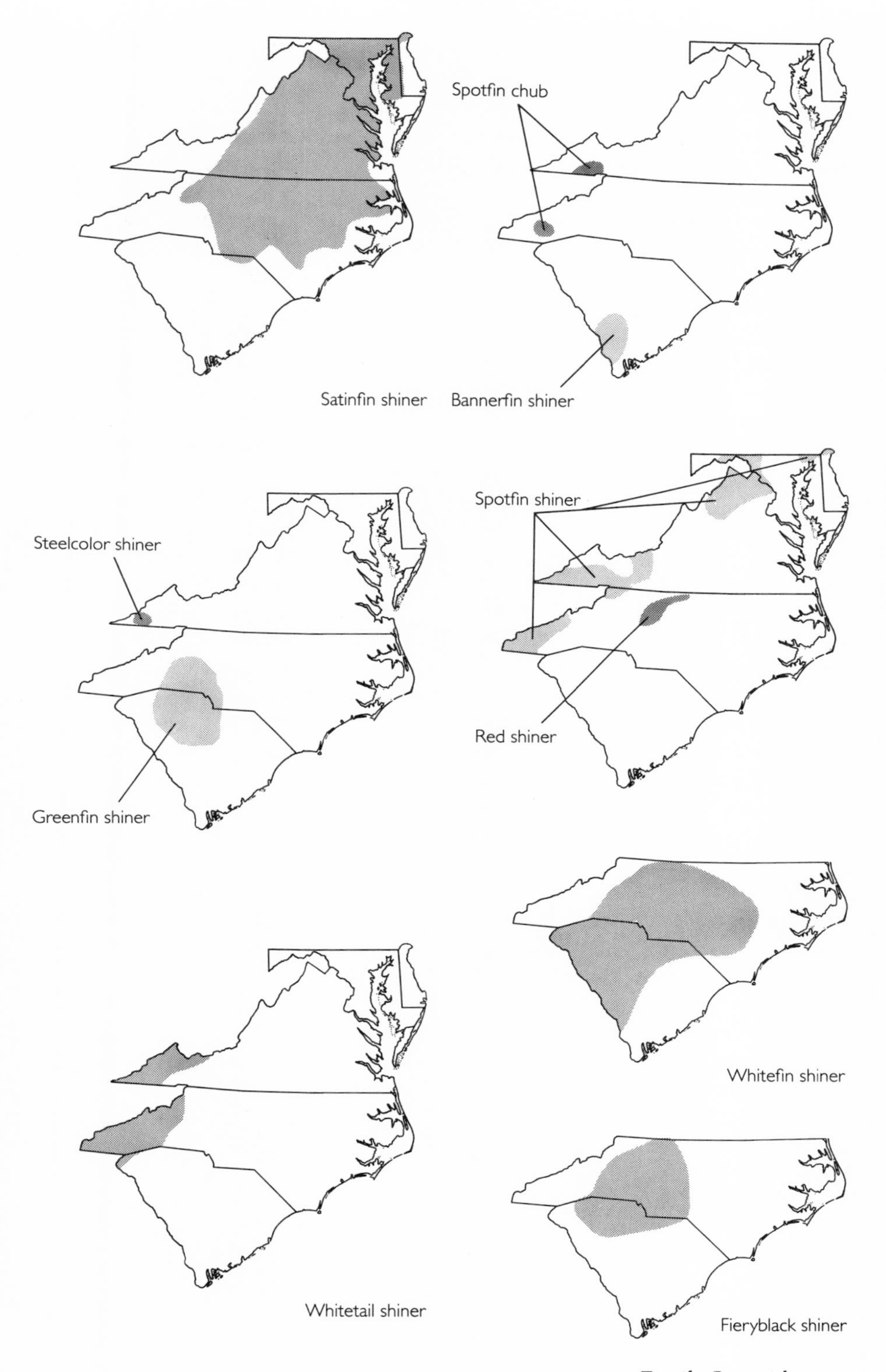
Spotfin chub
Satinfin shiner
Bannerfin shiner
Spotfin shiner
Steelcolor shiner
Red shiner
Greenfin shiner
Whitefin shiner
Whitetail shiner
Fieryblack shiner

feeders and feed on the surface, in midwater, and on the bottom. They favor aquatic insects, particularly larvae and nymphs of flies, gnats, and mayflies. Small fishes are occasionally eaten by the larger species.

Common carp

Cyprinus carpio Pl. 23

Description. 20.8 to 48.0 inches (528 to 1,220 mm). This is a robust gray-brown, large-scaled fish with one large barbel, preceded by a tiny barbel, on each side of the upper jaw. There is a strong sawtooth spine at the front of the dorsal and anal fins, and each one is preceded by two much smaller sawless spines. The body is strongly arched to the dorsal fin, followed by a gentle taper down to the tail, and it is flattened below. The dorsal fin is long, and the tail is forked. The mouth is located at the front of the head in the young and on the underside in the adult. The carp can attain a weight of 58 pounds, although large individuals are usually in the 5-to-15-pound range.

Many colorful and unusually scaled morphs, called nishikigoi by the Japanese and koi by us, have been produced through generations of selective breeding, much as in the goldfish, and there are now large breeders, shippers, and importers in various parts of the world, as a prize-winning koi can sell for tens of thousands of dollars.

Distribution and Abundance. The carp is native to temperate fresh waters in much of Europe and Asia. It was introduced into the United States in Boston in 1877, and by 1879 almost 13,000 individuals of this "wonder fish" had been distributed in 25 states and territories with the blessings and financial help of the U.S. Congress. It now occurs all across the country from the Atlantic to the Pacific, as well as across southern Canada and in much of northern Mexico. There are records from some of the large impounded rivers in central South Carolina; it is widely distributed in the other states of the region. It is often common and even abundant.

Habitat. The carp prefers still or slow waters with a muddy substrate and much vegetation, such as ponds, lakes, impoundments, and quiet portions of creeks and rivers. It often thrives in degraded areas where native fishes do not. Its grubbing in the substrate while searching for food results in the rooting out of much aquatic vegetation and in increased water turbidity, both of which are major causes of habitat degradation.

Natural History. The carp spawns in spring and early summer when the water temperature reaches the low 60s; spawning ceases once the water temperature exceeds the low 80s. As the water warms after winter, groups of carp move into shallow, warmer, vegetated waters and, with further warming, subdivide into smaller groups of one to three females and 2 to 15 males. The pursuing males and the subsequent spawnings result in much thrashing and

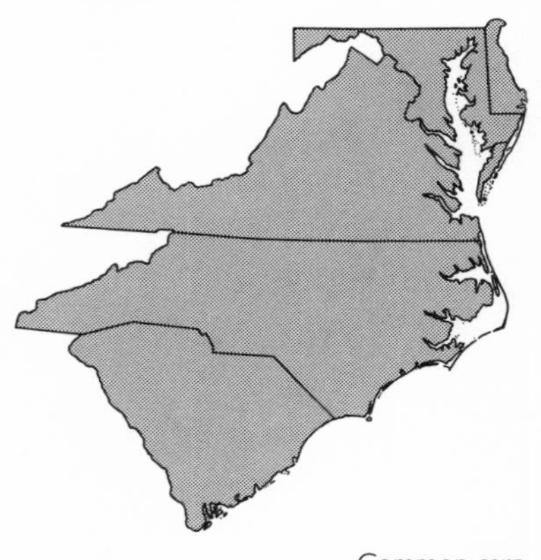

Common carp

splashing and sometimes jumping out of the water, which make this fish very conspicuous.

The numbers of eggs produced range from several tens of thousands to well over 2 million, depending on the size of the female. The tiny eggs adhere to submerged vegetation and to roots and hatch in from three to six days, varying with the water temperature. Growth depends on the temperature and on the availability of food, but the male matures when about three to four years old and the female when one year older. The maximum age attained in North American waters is just over 20 years.

The carp is an omnivore. It ingests mouthfuls of the soft bottom sediments (detritus) that are abundant in the eutrophic waters in which it is most common, expels them into the water, and then feeds on the disclosed insects, crustaceans, annelid worms, mollusks, weed and tree seeds, aquatic plants, and algae. The carp also feeds directly on food it finds on the bottom or on the water surface.

The carp has long been sought for protein in Eurasia, where it is often raised in farm ponds and fed boiled potatoes and other cheap foods. It was introduced into the United States for the same purpose. Europeans and Asians, who have historically had much less protein available in their diets than have Americans, eat it with gusto, usually when baked or smoked. But, probably because it is extremely bony and because of its mucky habitat, it is not a popular food fish in this country. Nonetheless, when smoked, the flesh is delicious and separates readily from the bones. Overall, however, its introduction was a huge commercial and ecological blunder. By the time this was first realized in the very early 1900s, this highly adaptive fish was already too widely distributed to root out, and we are now stuck with it. This large fish makes an exciting target for the archer when it forages or spawns in the shallows, and it is actively hunted in many parts of the region.

Silverjaw minnow

Ericymba buccata Pl. 24

Description. 1.7 to 3.8 inches (43 to 98 mm). This is a plain-colored minnow, tan above and silvery below. It has ten unique cavernous chambers of the lateral line system located on the lower cheeks and the side of the lower jaw, and these have the appearance of mother-of-pearl. The eye is large and directed upward, the under surface of the head is flattened, and the mouth is small, horizontal, and inferior.

Distribution and Abundance. The silverjaw minnow occurs within the mid-Atlantic region in western and northern Virginia and in the mainland of Maryland except its eastern portion. It occurs in much of the eastern United States and is often abundant.

Habitat. This minnow lives in clear and turbid creeks and rivers, where it is found in sandy-bottomed riffles and runs. It does best in areas of moderate gradient, and it is usually absent in cooler and high-gradient waters.

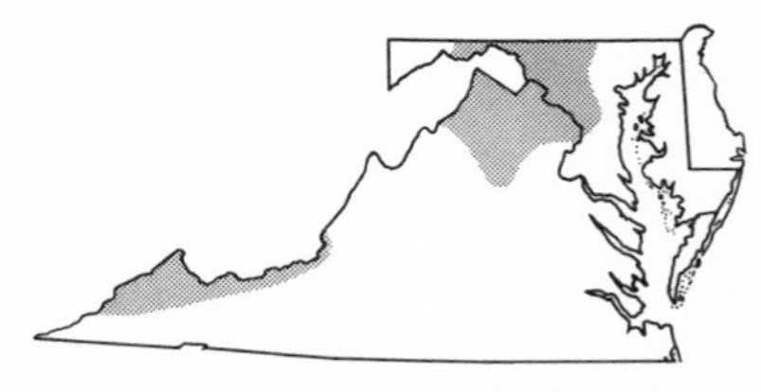

Silverjaw minnow

Natural History. The silverjaw minnow's food is young insects, small crustaceans, and detritus. It spawns from late March through early July, with a peak in early May and late June. The adults then school over fine gravel, and the female produces an average of 748 small eggs. Maturity is attained in one year. Most adults die after they spawn in their third year, but they can attain an age of up to four years. This fish is tolerant of some types of pollution but not of silt that has been deposited over sand or gravel. It is probably an important forage fish.

Blotched chub

Erimystax insignis Pl. 26

Description. 2.1 to 3.6 inches (54 to 92 mm). The blotched chub has a long and slender body that is deepest at the nape and flattened below. A small barbel is located at the corner of the subterminal mouth. There is an iridescent yellow stripe along the side, as well as seven to nine vertical dark gray to black blotches. A similar species, the **streamline chub**, *E. dissimilis* (Pl. 25), has 7 to 15 horizontal round dark blotches and a white to gold spot in front of, and another behind, the dorsal fin. Another species, the **slender chub**, *E. cahni*, has a series of dark ⟨-shaped marks on the rear half of the side.

Distribution and Abundance. The blotched chub occurs in the Tennessee River drainage of the mountains of North Carolina and Virginia. Outside the mid-Atlantic region it occurs at lower elevations further west in this drainage and in the Cumberland River drainage. It is generally uncommon but is sometimes common locally. The streamline chub in the region occurs only in extreme western Virginia and adjacent northwestern

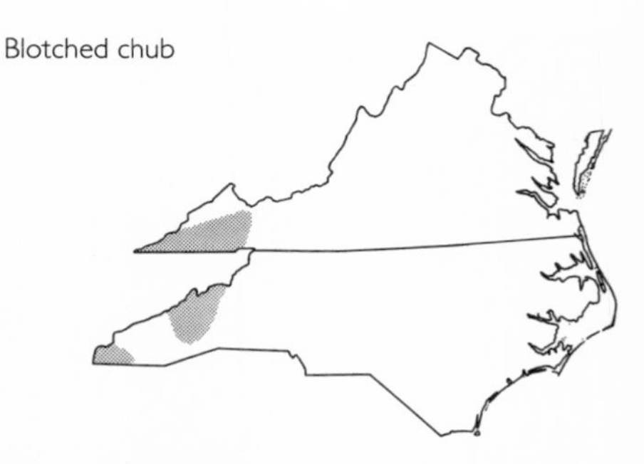

North Carolina. The slender chub occurs only in southwestern Virginia within the region, and it is listed as threatened by the federal government.

Habitat. The blotched chub is typically found over coarse gravel in riffles and runs of small to medium-sized rivers with a strong current.

Natural History. The biology of the blotched chub is poorly known. It spawns in late spring to early summer. The food is filamentous algae and a variety of bottom invertebrates that it finds with its taste buds, rather than its eyes, as it gropes on riffle bottoms.

Cutlips minnow

Exoglossum maxillingua Pl. 27

Description. 3.9 to 6.2 inches (100 to 157 mm). The cutlips minnow is nondescript except for its highly specialized lower jaw, with its three distinct lobes: two thick and fleshy lateral lobes and a larger, thinner, and bony central one. The body is stout (not compressed) and the color is mostly somber: olive to gray above and paler on the belly; often the side has a violet hue. It lacks a barbel at the corner of the mouth. In the closely related **tonguetied minnow**, *E. laurae*, the lower jaw is bordered by a pair of low flaps, and a small barbel is present near the corner of the mouth.

Distribution and Abundance. The cutlips minnow is found in upland areas from Canada and New York to north central North Carolina. It is often common, but it is considered endangered in North Carolina because of its limited distribution there. The tonguetied minnow occurs in the mid-Atlantic region only in the New River drainage of the mountains of North Carolina and Virginia.

Habitat. The cutlips minnow is found in small to moderate-sized clear creeks and small rivers with a gravel, rubble, and boulder bottom and little vegetation. There it is usually found in quiet waters near the bottom of pools and under undercut banks.

Natural History. The cutlips minnow breeds in May and June. The male gathers stones from a distance of several feet and constructs a pebble-mound nest, which may be up to three inches high and up to 18 inches in diameter. He establishes a position at the crest of the nest to attract females. Spawning occurs on the upstream slope of the nest, and the eggs are deposited in the gravel. The male remains on the nest after breeding and keeps it clear of silt. The young remain in the nest for several days after they hatch. The cutlips minnow feeds on bottom-living insects and mollusks and on algae and detritus. It is known to bite out the eyes of other fishes, particularly when confined with them in aquariums and bait buckets.

Cutlips minnow

Eastern silvery minnow

Hybognathus regius Pl. 28

Description. 2.2 to 4.7 inches (55 to 120 mm). The eastern silvery minnow can be mistaken for one of the shiners but it is distinguished by a small, slightly subterminal, crescent-shaped mouth, a long and coiled intestine, and a black peritoneum. The body is usually silvery, but a large male may have a yellow tinge on the side.

Distribution and Abundance. This fish is found on the Atlantic slope from Canada to Georgia. It is widely distributed in the mid-Atlantic region and is generally common.

Habitat. The silvery minnow lives in slow larger creeks and rivers and prefers pools and backwaters.

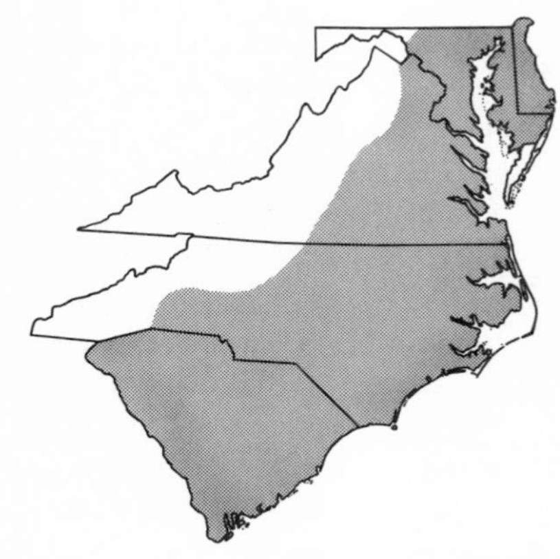

Eastern silvery minnow

Natural History. This fish reproduces in late spring in quiet backwaters, usually over vegetation. Two or more males spawn with a single female in water only about six inches deep, and a female can lay up to 6,600 eggs. The eggs sink (are demersal) but they do not adhere to the bottom. A few fish may mature by age one, but most mature in the following year. The primary food is bottom ooze and algae, and the long intestine is an adaptation to digest it.

Bigeye chub
Hybopsis amblops
Highback chub
Hybopsis hypsinotus Pl. 29
Santee chub
Hybopsis zanema Pl. 32

Description. Five species of *Hybopsis* occur in the region. They resemble shiners but have a modest barbel at the corner of the mouth, a long snout, an inferior mouth, and upward-looking eyes.

Bigeye chub: 2.0 to 3.5 inches (51 to 90 mm). This species is identified by a prominent black stripe on the side. As the name implies, the eye is large, and it is as wide as the snout is long. The body color is light yellow. The **rosyface chub**, *H. rubrifrons* (Pl. 31), is similar, but the eye is smaller and the breeding male develops red on the anterior one-third of the body.

Highback chub: 1.7 to 2.8 inches (43 to 72 mm). The body is strongly arched, and this fish has a highback appearance. The body is dark olive above, with a purplish or black lateral stripe that extends onto the snout. The fins of a breeding male are red.

Santee chub: 1.8 to 3.0 inches (46 to 75 mm). This chub has a long snout, a slender body with dark cross-hatching on the back and side, and a moderate barbel. The breeding male is silvery, with yellow fins, and dusky black streaks are present in the dorsal fin. The **thicklip chub**, *H. labrosa* (Pl. 30), is highly similar, but it has small dark brown blotches on the back and side.

Distribution and Abundance. In the mid-Atlantic region, the bigeye chub is found only in the mountains of Virginia and North Carolina. It is widely distributed in the east central United States from western New York to the Ozark Mountains. The highback chub is restricted to the piedmont and Blue Ridge foothills in the Pee Dee and Santee drainages of Virginia and the Carolinas. The range of the Santee chub generally overlaps that of the highback chub, but a few populations of the former are also found in the coastal plain of both Carolinas. These populations, however, may be an undescribed species, and those in North Carolina are listed as of special concern. None of the three species is usually encountered, but any of them can be locally common.

Habitat. The bigeye chub prefers clear creeks and small rivers with a sand bottom, and it is rarely associated with aquatic vegetation. It appears to be highly vulnerable to siltation and other types of pollution. Highback and Santee chubs inhabit small creeks to medium-sized rivers with a sand or rock bottom and can tolerate more turbid and warmer waters than can the bigeye chub.

Natural History. These species spawn from late spring to mid-summer. Food of the bigeye chub is aquatic insects, and its large eye suggests sight feeding rather than a reliance on taste buds located on the barbels.

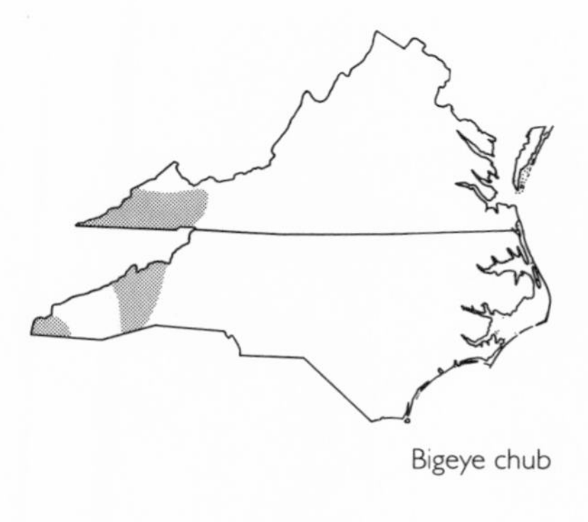
Bigeye chub

Thicklip chub

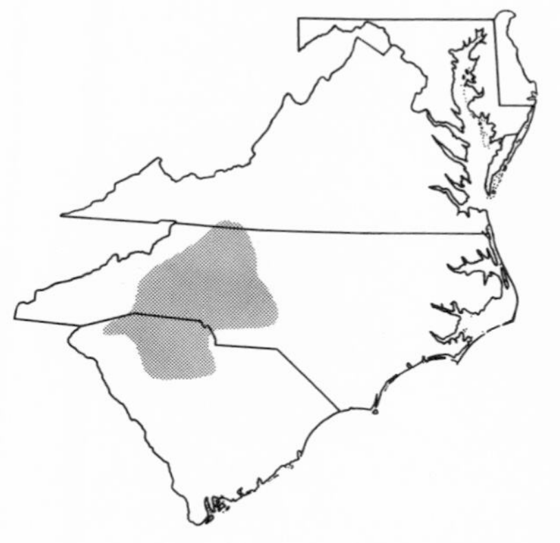
Highback chub

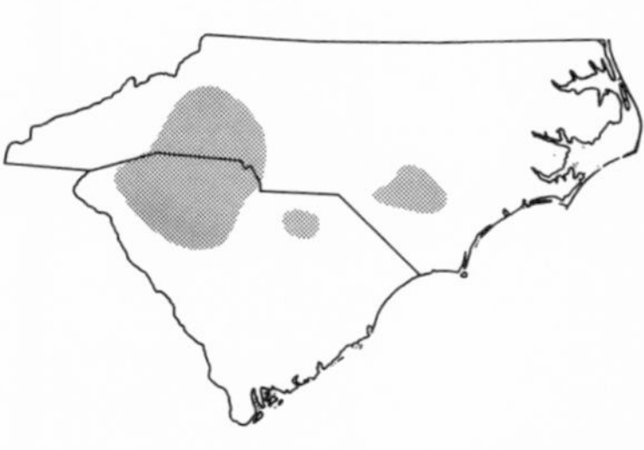
Santee chub

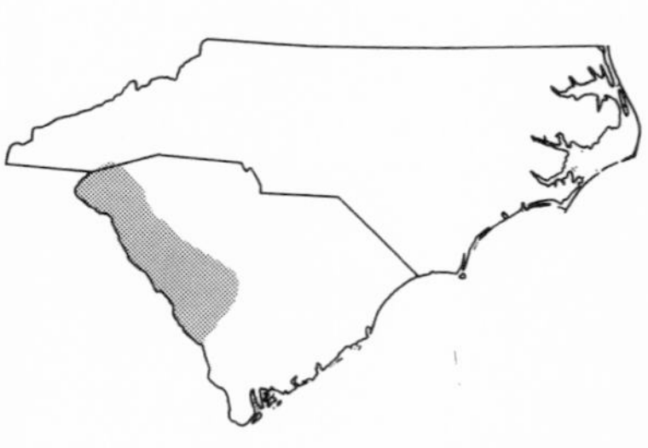
Rosyface chub

White shiner
Luxilus albeolus Pl. 33
Crescent shiner
Luxilus cerasinus Pl. 34
Warpaint shiner
Luxilus coccogenis Pl. 35
Common shiner
Luxilus cornutus Pl. 36

Description. Five species of *Luxilus* occur in our area. All have a strongly compressed and deep body with large lateral scales that are much higher than wide.

White shiner: 3.0 to 6.1 inches (75 to 156 mm). This fish has a silvery body, and there is usually a silvery patch on the gill cover of the adult. The side lacks dark scales or has only a few, and red color does not develop on the body of the breeding male.

Crescent shiner: 2.7 to 4.5 inches (69 to 114 mm). Irregular blackened scales on the side form crescent-shaped bars, and there is much red on the body and fins of the breeding male.

Warpaint shiner: 3.0 to 5.9 inches (76 to 151 mm). There is a prominent red bar on the opercle, a black bar or slash on the side behind the opercle, and a light base and a dark stripe near the edge of the caudal and dorsal fins. In the breeding male the snout and upper lip are red and the lower fins white.

Common shiner: 3.3 to 8.2 inches (83 to 208 mm). This species is similar to the white shiner, but it has one or two dark parallel stripes on the upper side and several to many dark crescents (blackened scales) on the side. The **striped shiner**, *L. chrysocephalus*, is similar to the common shiner, but it has four dark stripes on the upper body.

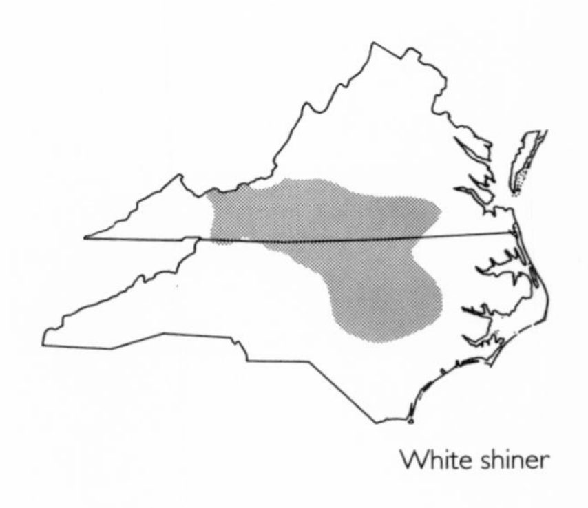
White shiner

Distribution and Abundance. The common shiner in the the mid-Atlantic region occurs from Delaware south to Virginia; it is widely distributed throughout the upper half of the eastern United States. The other species are more restricted: the warpaint shiner occurs only in the upper Tennessee River drainage in the mountains of Virginia and the Carolinas, and the other two are generally sympatric (have overlapping ranges) in Virginia and North Carolina. All are often locally common.

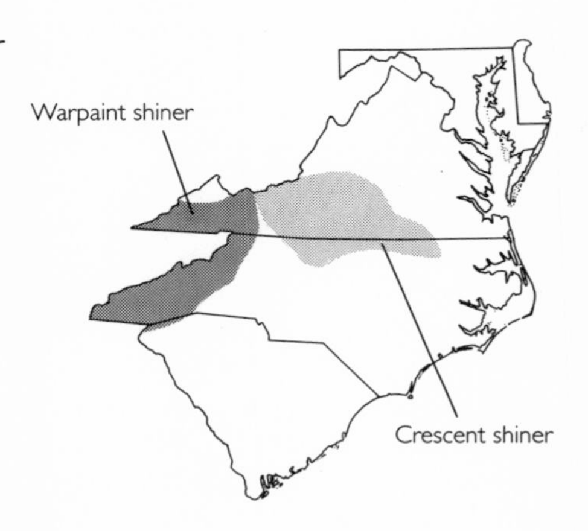

Habitat. These fishes prefer flowing pools in medium-sized creeks and small rivers with a gravel or rock bottom. The white shiner tolerates turbid waters better than the others.

Natural History. These species spawn from mid-May to mid-July, depending on the water temperature, over nests built by other minnows, particularly chubs of the genus *Nocomis*, or over gravel in riffles. The male arrives first in the spawning area and attempts to hold a territory. The female moves upstream into the male's territory when she is ready and then can lay up to 4,000 eggs. Most individuals are mature by age two and generally live up to four years, and the common shiner can attain an age of six years. The common shiner is an omnivorous and opportunistic feeder on plant and animal matter. The warpaint shiner feeds extensively at the surface on terrestrial insects, while the other shiners feed primarily on bottom-dwelling insect nymphs and larvae.

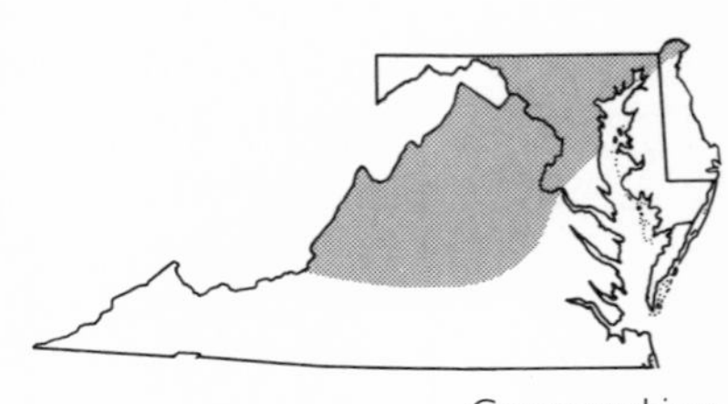
Common shiner

Rosefin shiner
Lythrurus ardens Pl. 37
Mountain shiner
Lythrurus lirus
Pinewoods shiner
Lythrurus matutinus Pl. 38

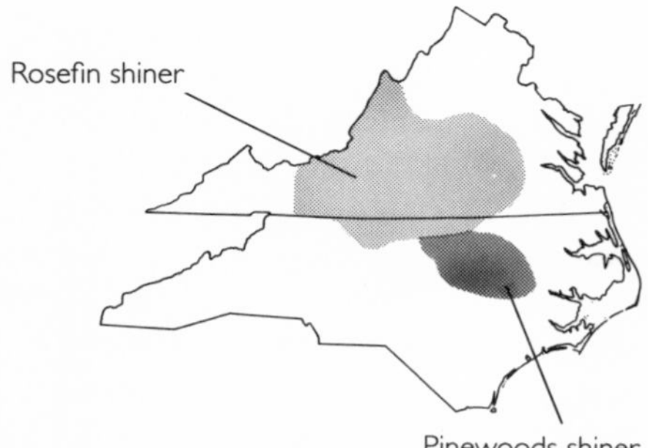

Description. 1.8 to 3.5 inches (46 to 89 mm). All three species have an elongate and slim body. The rosefin shiner is slightly compressed, and the scales before the dorsal fin are tiny. There is a dark blotch in the dorsal fin at the origin. On the breeding male the fins become red, the head and side become orange-red, and the dorsum becomes steel blue, and there are dusky bars on the back that extend down on the side. The pinewoods shiner appears similar, but it has a more slender and blue-gray body, a whitish venter, and red on most of the interior of the unpaired fins. It also has a black slash behind the opercle. The breeding male develops a bright red head, with a rough-textured white color on top due to the numerous small breeding tubercles then present. The mountain shiner is slender, and it lacks the blotch in the dorsal fin and the bright red breeding color of the male found in the other two species.

Distribution and Abundance. The rosefin shiner has two centers of occurrence: the Atlantic slope from the York River in Virginia south to the extreme upper Cape Fear River in North Carolina, and (outside of the mid-Atlantic region) in the central United States from Ohio to Alabama. The pinewoods shiner is restricted to the Tar and Neuse drainages of North Carolina. The mountain shiner occurs in the mid-Atlantic region only in extreme southwestern Virginia. All can be common to abundant.

Habitat. Rosefin and mountain shiners occur in the midwater area of pools and runs of clear, moderately swift, rocky creeks and small rivers. The pinewoods shiner prefers the midwater area in sandy runs and pools in creeks and small rivers.

Natural History. Rosefin and pinewoods shiners spawn in spring and early summer, often over nests of chubs of the genus *Nocomis*, or in fast riffles. The male is larger than the female and maintains a territory over the spawning area. Maturity is reached after the first year. Insects, principally terrestrial species, are the main food and are presumably taken from mid-depth to the surface, although aquatic insects and algae are also eaten. The rosefin shiner does not tolerate siltation.

Pearl dace
Margariscus margarita Pl. 39

Description. 2.5 to 6.3 inches (65 to 160 mm). This species is not distinctive and is often confused with others. It has a nearly cylindrical body, a rounded snout, a small, slightly subterminal mouth, small scales, a short head, and usually a flaplike barbel in the groove above the mouth near the rear of the upper jaw. The back is olive and the venter whitish, there is a black stripe at midbody, and there are many small black and brown specks on the silvery side. The breeding male develops a bright orange-red stripe below the lateral black stripe and many small tubercles on the head.

Distribution and Abundance. The pearl dace occurs in the region in the northern portion of Virginia and the mainland of Maryland except its southeastern corner; it is widely distributed in the northern United States and Canada. It is uncommon to locally common.

Habitat. The pearl dace occurs in a wide range of cold to cool and clear to slightly turbid, unpolluted creeks, especially their headwaters, and rarely in ponds, lakes, and rivers. It occurs most frequently over sand and gravel and less commonly over silt, mud, rubble, and boulders. It often lives among vegetation.

Natural History. The pearl dace spawns from March to June, with the exact time depending on geographic location and annual temperature variation. Spawning has been observed at a water temperature of 63° to 65°, in a strong current to nearly quiet water, and at a water depth of 1½ to 4½ feet. It builds no nest. Territories are about eight inches wide and at least six feet apart. An invading male is usually pursued out of the territory, and a ripe female is often driven into it. Spawning lasts for about two seconds and occurs as the male cups his pectoral and tail fins around the female. The female spawns with many different males, and presumably only a few eggs are released at one time. One female was reported to contain 1,686 eggs. Maturity is probably attained in one year in some cases, but it usually takes two. It is a rare male that survives one spawning season, while a few females can attain an age of four years. Food is primarily insects, but it also includes phytoplankton, mollusks, water mites, algae, vegetable detritus, and small fishes. It is thought to be a moderately important forage fish.

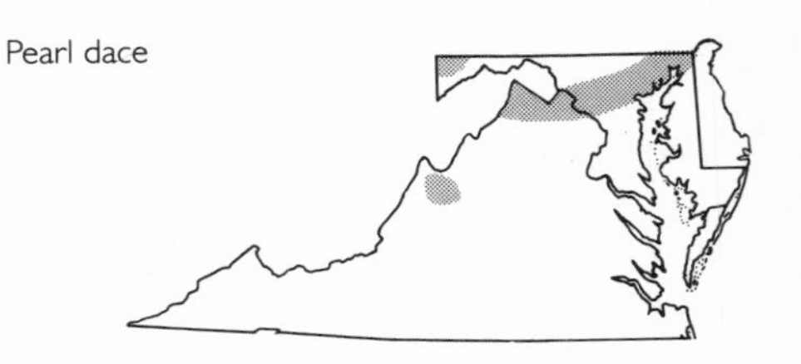

Bluehead chub
Nocomis leptocephalus Pl. 40
River chub
Nocomis micropogon Pl. 41
Bull chub
Nocomis raneyi

These are large, bronze-colored, and stout-bodied minnows. The scales on the back and side are edged with black and give the side a weak cross-hatched appearance, the mouth is large and subterminal, and there is a small barbel in the corner of the mouth. The genus occurs in the eastern United States and adjacent Canada, and three additional species are found outside the mid-Atlantic region.

Description. Bluehead chub: 2.9 to 10.2 inches (75 to 260 mm). This fish has a short and rounded snout, a stout and light brown body, and light yellow to red-orange fins. The breeding male has a blue head, an orange or blue side, orange fins, a large hump on top of the head, and prominent breeding tubercles on top of the head and behind the level of the nostrils.

River chub: 3.6 to 12.6 inches (92 to 320 mm). The river chub has a longer snout than the bluehead chub. The breeding male has a head and lower body that are pink-blue, some fins that are red or orange, a hump on top of the head, and tubercles on the side as well as on the front of the snout and just before the nostrils. A similar species, the **bigmouth chub**, *N. platyrhynchus*, differs from it by tubercles that occur from the snout to over the top of the head and to well behind the eye.

Bull chub: 4.7 to 12.6 inches (120 to 320 mm). This is the largest species of *Nocomis*, and it is nearly identical to the river chub but has a smaller mouth and tubercles that occur back to between the eyes. Use the distribution of these two chubs as described below to help identify them.

Distribution and Abundance. The bluehead chub in the mid-Atlantic region occurs on the Atlantic slope of Virginia and the Carolinas but it is absent from most of the coastal plain. The river chub is found in the mountains of the Carolinas, the northern two-thirds of Virginia (but excluding most of the coast), and Maryland west of Chesapeake Bay (but it is absent from Calvert County south). Both species are also more widely distributed in the eastern United States. The bull chub occurs only on the Atlantic slope of southern Virginia and north central North Carolina. The bigmouth chub is found only in the New River drainage of North Carolina, Virginia, and West Virginia. All four species are often common and sometimes locally abundant.

Habitat. These are fishes of clear creeks and small rivers with a medium to fast current and a substrate of rock, gravel, or sand. The young, especially, may occur in riffles that are only a few inches deep, while older fishes are usually found in deeper waters.

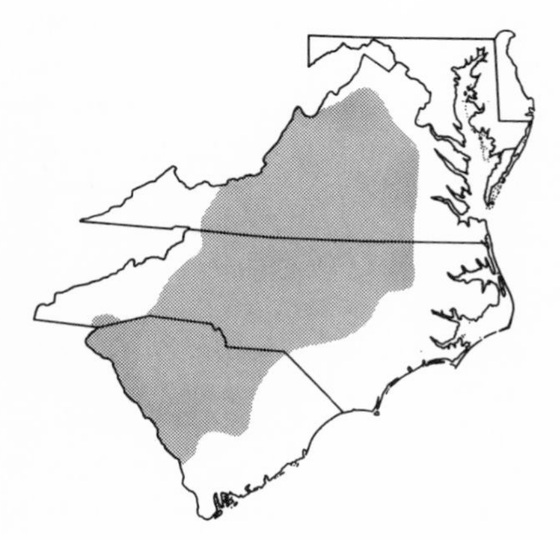

Bluehead chub

River chub

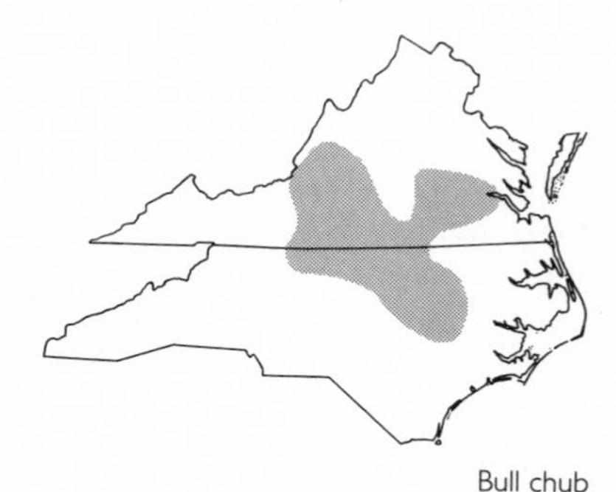

Bull chub

Natural History. All four species feed on insects, small crustaceans, crayfishes, snails, fishes, and plants. The bluehead chub eats more plants than do the other three chubs, and its intestine is longer than that of most chubs. The lips are well supplied with taste buds. The male of all four species carries gravel and stones in his mouth and builds a gravel mound nest on which spawning occurs. River chub nest mounds have been reported to be 33 to 43 inches in diameter and six to seven inches high; they are oval or round. One mound was comprised of some 7,000 pebbles that weighed a total of 235 pounds and had a volume of 70.5 quarts. Nests are usually located in water about a foot deep, and less commonly in water up to three feet deep. The river chub takes 20 to 30 hours to prepare a nest. The eggs are deposited in a small trench dug by the male in the top of the mound. The female bluehead chub has been noted to contain 710 to 800 eggs, and the female river chub 400 to 625.

At spawning the male clasps the receptive female between the upper surface of one of his pectoral fins and the side of his body. The eggs and milt are then simultaneously extruded, the female is released, and the male covers the trench. Nest building and spawning usually occur in May and June when the water temperature is in the range of 59° to 68°. Maturity in these four species is attained when the fishes are two or three years old, and they can reach an age of four or five years. The male chub tolerates the use of his nest by many other species of fishes, and this tolerance probably explains the relatively large number of intergeneric hybrids that have been reported for *Nocomis.*

Golden shiner

Notemigonus crysoleucas Pl. 42

Description. 2.5 to 11.8 inches (64 to 300 mm). This silver to brass-colored minnow is characterized by a slab-sided and rhomboid-shaped body, a small rhomboid-shaped head, and falciform (sickle-shaped) fins. The lateral line is conspicuous, extends the length of the body, and, unlike in other fishes of the mid-Atlantic region, dips down to near the lower edge of the side instead of being located at midbody level. The midline of the belly from the pelvic fins to the anus is naked (lacks scales), and it has a sharp keel.

Distribution and Abundance. The golden shiner is found in fresh water in much of the region. It is abundant to common in the coastal plain and piedmont, and it is uncommon in the mountains. More generally, it occurs in the eastern half of the United States from Maine to Florida and Montana to Texas and in adjacent Canada.

Habitat. This species is typically found in slow, vegetated, and soft-bottomed waters such as lakes, oxbows, ponds, canals, swamps, and quiet-water portions of creeks and river edges. It is an active fish and swims in a loose school off the bottom.

Golden shiner

Natural History. The golden shiner begins to spawn when the water temperature reaches the high 60s and continues to the late summer. The eggs are adherent and are scattered, usually over filamentous algae or rooted aquatic plants, and abandoned. The young golden shiner can reach a length of three inches in its first summer, and presumably it can reproduce while still in its first year of life. Food is primarily adults and immature stages of cladocerans, midges, dragonflies, beetles, and water mites, as well as filamentous algae and mollusks. The food of the young and adult is apparently similar. This shiner feeds from midwater to the surface. It in turn is eaten by predatory fishes such as basses, pickerels, and sunfishes as well as by herons, egrets, kingfishers, and water snakes.

The golden shiner is a highly favored bait fish, often sold in bait and tackle shops. Such fishes may have been caught in the wild or raised in ponds. The golden shiner is also pond raised as food for pond-cultured largemouth bass. Bait bucket releases and others have resulted in its establishment in many areas outside of its natural range.

Ironcolor shiner
Notropis chalybaeus Pl. 44
Redlip shiner
Notropis chiliticus Pl. 45
Dusky shiner
Notropis cummingsae Pl. 47
Tennessee shiner
Notropis leuciodus Pl. 49
Yellowfin shiner
Notropis lutipinnis Pls. 50–51
Taillight shiner
Notropis maculatus Pl. 52
Coastal shiner
Notropis petersoni Pl. 54
Silver shiner
Notropis photogenis Pl. 55
Swallowtail shiner
Notropis procne Pl. 56
Saffron shiner
Notropis rubricroceus Pl. 58
New River shiner
Notropis scabriceps Pl. 59
Sandbar shiner
Notropis scepticus Pl. 60
Mirror shiner
Notropis spectrunculus Pl. 61

Notropis contains 71 species and is the second-largest genus (after the darters, genus *Etheostoma*) of freshwater fishes in North America. With so many species, it is difficult to diagnose the genus easily, but a number of characters, which overlap with those of other genera of minnows, that will identify it are eight dorsal fin rays, scales on the nape of about the same size as those on the upper side, scales on the front half of the side not much taller than wide, a short intestine, and a lack of barbels. Twenty-eight species of shiners occur in the mid-Atlantic region. Many are widespread in the region and even throughout eastern North America, but some are highly localized. One of the latter, the Cape Fear shiner, occurs in only a small portion of the Cape Fear River drainage in central North Carolina, and it is listed as endangered by the federal government.

Description. Ironcolor shiner: 1.8 to 2.6 inches (45 to 65 mm). This shiner is distinguished by the combined presence of a well-defined black lateral stripe on the side from the caudal fin base to around the snout (with parts of the stripe on both the chin and the lips), a snout that is shorter than the eye is wide, a back and upper side that are straw yellow, a mouth that is small, and a roof of the mouth that is black.

Redlip shiner: 1.5 to 2.8 inches (38 to 72 mm). The redlip shiner has bright red lips, red coloration in the dorsal, anal, and caudal fins, and scattered black blotches on the side. The breeding male also has a scarlet-red body and eye and yellow fins.

Dusky shiner: 1.4 to 2.8 inches (36 to 72 mm). This fish differs from the similar ironcolor shiner by having a more broad dark lateral stripe, which extends down to below the lateral line and rearward onto the caudal fin, and a bright orange stripe above the dark. The similar **highfin shiner**, *N. altipinnis* (Pl. 43), differs from it by having a more narrow and more diffuse dark lateral stripe that does not extend below the lateral line and a yellow stripe above the dark one.

Tennessee shiner: 1.7 to 3.2 inches (43 to 82 mm). This is a silvery-sided shiner that is identified by black dashes along the lateral line and a small black rectangle at the base of the caudal fin. A bright green stripe, best viewed when the fish is in the water, is present on the upper side. There are dark wavy lines on the back and upper side. The body and fins of the breeding male are suffused with red. The similar **telescope shiner**, *N. telescopus* (Pl. 62), lacks a black rectangle at the base of the caudal fin, has a larger eye, and has more distinct wavy lines on the back. The **popeye shiner**, *N. ariommus,* is also similar but it has a very large eye, and it lacks wavy lines on the back and dashes along the lateral line.

Yellowfin shiner: 1.9 to 3.0 inches (48 to 75 mm). This robust shiner has a reddish-brown back and upper side and a dark lateral stripe. The breeding male develops a bright pink-red color on the body and bright lemon yellow to red fins. The **greenhead shiner**, *N. chlorocephalus* (Pl. 46), is similar, but the breeding male is red with bright white fins.

Taillight shiner: 2.3 to 3.0 inches (58 to 76 mm). This slender shiner with large, pointed fins is easily distinguished by a large (larger than the pupil of the eye) black spot at the base of the caudal fin and by a black blotch on the front of the dorsal fin. The breeding male develops a bright red color on the body and on the distal edge of the fins.

Coastal shiner: 1.9 to 3.2 inches (48 to 82 mm). This fish has a black lateral stripe that ends in a wedge-shaped spot at the base of the caudal fin. All rays of the anal fin are edged with black. It has a relatively long snout that slightly overhangs the mouth. The similar **spottail shiner**, *N. hudsonius* (Pl. 48), has a large eye, a short and rounded snout, and a large caudal fin spot.

Silver shiner: 3.0 to 5.4 inches (77 to 137 mm). This large and silvery minnow has a slender and compressed body, a large and terminal mouth on a long snout, and two black crescents between the nostrils. Four other shiners are similar in appearance. The **emerald shiner**, *N. atherinoides,* lacks black crescents between the nostrils and has a smaller eye and a shorter snout. The **rosyface shiner**, *N. rubellus* (Pl. 57), also lacks crescents but has a sharper snout, red coloration on the head and body of the male, and a black streak above the silver lateral stripe. The **comely shiner**, *N. amoenus,* has crescents between the nostrils, but it has a smaller eye than the silver shiner and is found only in Atlantic slope drainages. The **roughhead shiner**, *N. semperasper,* has a black stripe on the side and is restricted to the upper James River drainage in Virginia.

Swallowtail shiner: 1.4 to 2.8 inches (36 to 72 mm). Small and delicate looking, the swallowtail shiner is characterized by black dashes along the lateral line and a dark lateral stripe that does not continue around the snout. The breeding male develops a yellow body and fins. Five other shiners are similar in appearance. The **whitemouth shiner**, *N. alborus*, has a jagged-edged stripe on the side, which continues around on the snout or upper lip. The **bridle shiner**, *N. bifrenatus*, has an incomplete lateral line and a lateral stripe that continues around on the upper lip. The **sand shiner**, *N. ludibundus*, has a fine (very narrow) dusky stripe on the back that expands into a dark wedge at the dorsal fin origin. The **Cape Fear shiner**, *N. mekistocholas* (Pl. 53), also has a dark stripe on the side, and it is the only shiner in the region with a long, coiled dark gut that is visible through the wall of the belly. The **mimic shiner**, *N. volucellus*, has scales on the side that are much deeper than wide and a broad, rounded snout.

Saffron shiner: 1.9 to 3.3 inches (48 to 84 mm). Similar to the redlip shiner, the saffron shiner has on the side a black stripe that starts behind the head and a clear spot on the body at the posterior end of the dorsal fin insertion.

New River shiner: 1.9 to 3.3 inches (48 to 84 mm). A broad snout and large and upward-directed eyes separate this shiner from others. Its mouth is slightly subterminal, and it has black dashes along the lateral line. The side is silvery, and there is a dusky lateral stripe.

Sandbar shiner: 2.4 to 3.5 inches (60 to 90 mm). This shiner is identified by a silvery side with a dusky lateral stripe, a large round eye that is wider than the snout is long, scales that are darkly outlined on the back, a concave anal fin, and black dashes along the lateral line.

Mirror shiner: 1.7 to 3.0 inches (43 to 75 mm). This shiner is identified by a slender body, a broad head with a rounded snout, the absence of scales on the nape, a black wedge-shaped caudal spot, a small subterminal mouth, and upward-directed eyes. The breeding male has red-orange fins with a white edge. An undescribed shiner, the **sawfin shiner**, is similar but has black specks along the first four dorsal fin rays rather than along all the rays as in the mirror shiner.

Distribution and Abundance. The ranges of the 13 species discussed fall into three categories, with some overlap: (1) the silver shiner, New River shiner, saffron shiner, Tennessee shiner, and mirror shiner occur in the mountains of the region, while (2) the sandbar shiner, yellowfin shiner, redlip shiner, and swallowtail shiner generally occur in the piedmont, and (3) the coastal shiner, ironcolor shiner, dusky shiner, and taillight shiner are common on the coastal plain. Some species, such as the New River and the yellowfin shiners, are restricted to a small geographic area, while others, such as the ironcolor and the swallowtail shiners, have a large range. All of the shiners can be locally common to abundant. Because of its small range in North Carolina, the yellowfin shiner is there listed as of special concern.

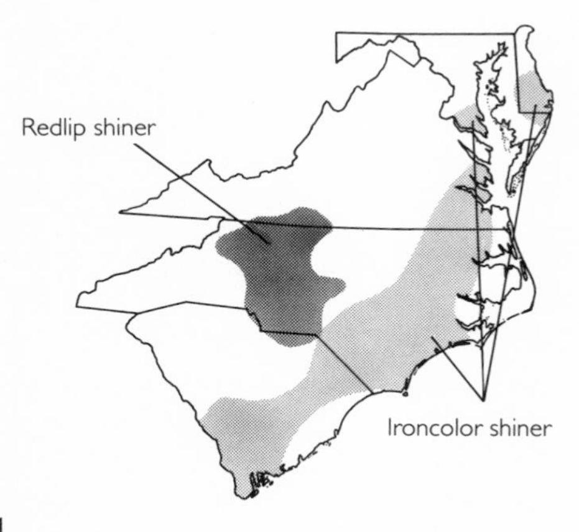

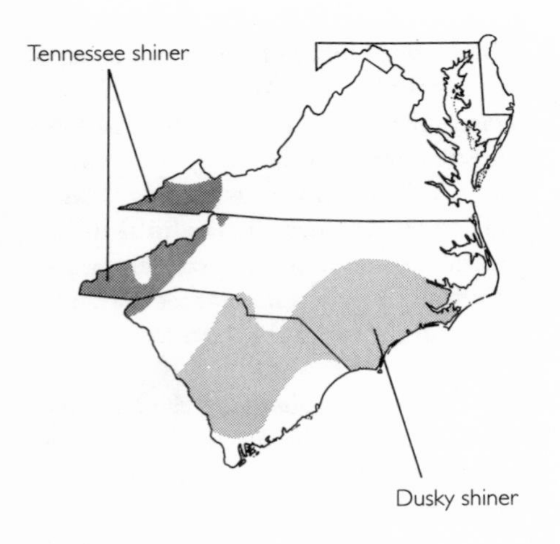
Tennessee shiner
Dusky shiner

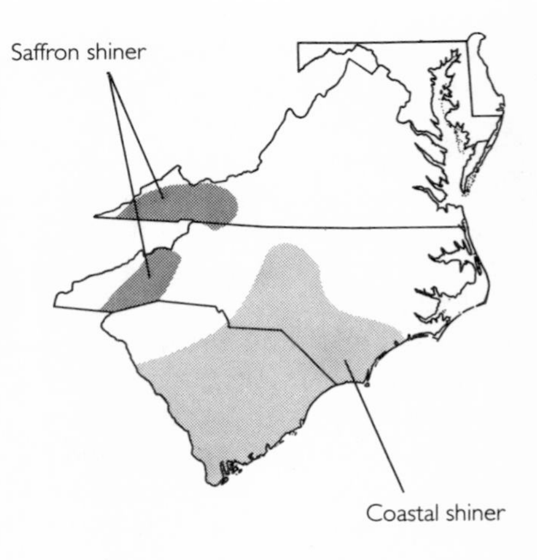
Saffron shiner
Coastal shiner

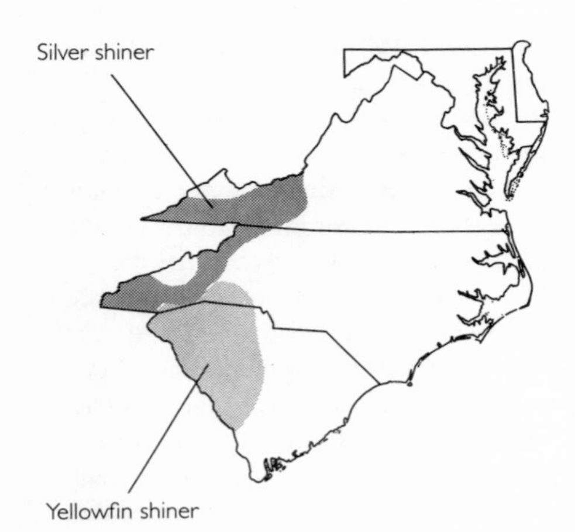
Silver shiner
Yellowfin shiner

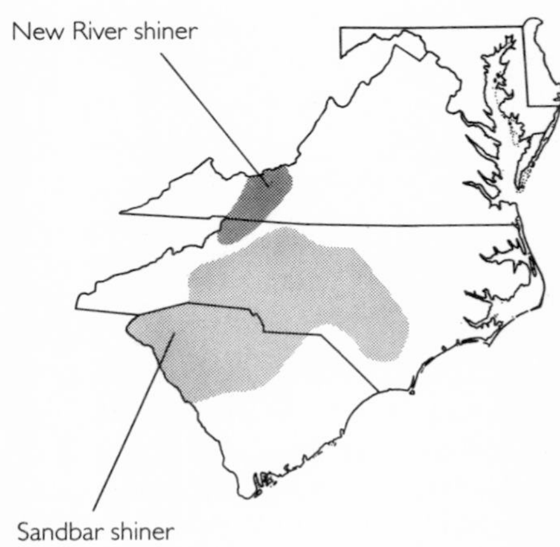
New River shiner
Sandbar shiner

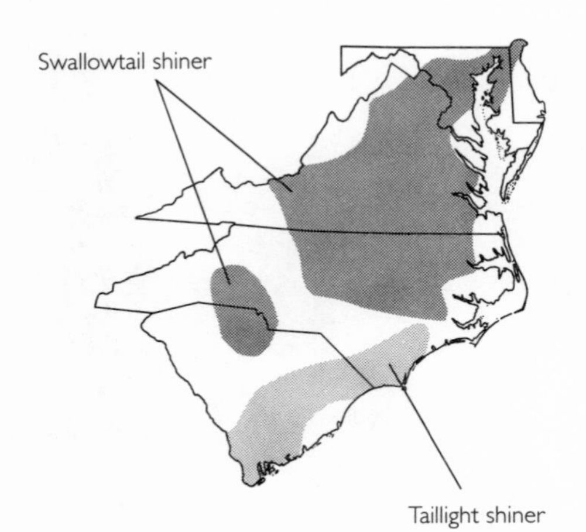
Swallowtail shiner
Taillight shiner

Mirror shiner

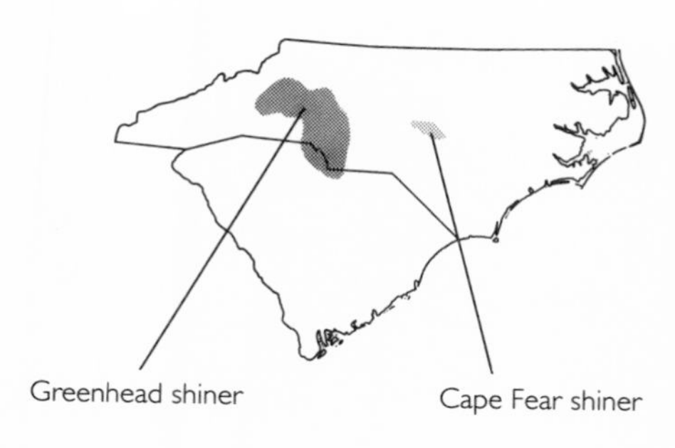

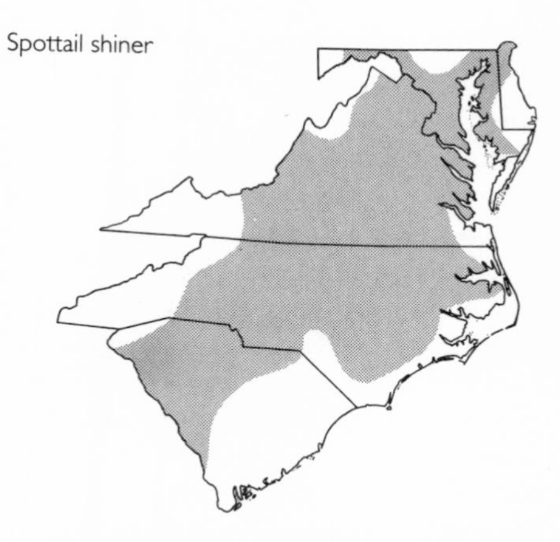

Habitat. Shiners generally occur from headwater creeks to small rivers in pools and runs with a sand or rock bottom. The coastal and taillight shiners also occur in still waters of lakes.

Natural History. Although these shiners are often numerous in streams, as well as in museum collections, little is known about many of the species. Most presumably spawn in the spring and can live up to three years. Food mostly consists of larvae of aquatic and terrestrial insects, supplemented with diatoms, algae, and small crustaceans.

Pugnose minnow

Opsopoeodus emiliae Pl. 63

Description. 1.4 to 2.5 inches (36 to 64 mm). This is a delicate-looking minnow with a distinctive small, nearly vertical mouth and a short, blunt snout. The scales of the back and the upper portion, or all, of the side are edged with black and create a cross-hatched pattern. The back and upper side is dusky yellow, the lower side and belly is silvery, and there is a dark stripe on the side from the head to the base of the tail. The breeding male develops a dark silver-blue body, and the normally dusky dorsal fin becomes blackish on the anterior and posterior portions, accentuating a white area in the center of the fin. The tips of the anal and pelvic fins become bright white. Small white knobs develop on the first three rays of the dorsal fin of a reproductive male. This species has often been included in the shiner genus *Notropis*, but its current placement is probably more accurate.

Distribution and Abundance. In the mid-Atlantic region this minnow is found only in the southern portion of South Carolina, from the Edisto River southward. It also occurs in much of the southeastern United States well into Texas and in the Mississippi River basin north to central Wisconsin and eastern Ohio. This species is common to rare depending on location, and it has been greatly reduced in numbers and even extirpated in some parts of its range.

Habitat. The pugnose minnow occurs in a wide range of waters, from small creeks, sloughs, borrow pits, and lakes, to large rivers. It prefers clear, quiet, heavily vegetated water over a soft (mud) bottom.

Natural History. This species feeds on midge larvae, algae, and tiny crustaceans. In the mid-Atlantic region it probably breeds from April to September. Many small, crowded spawning tu-

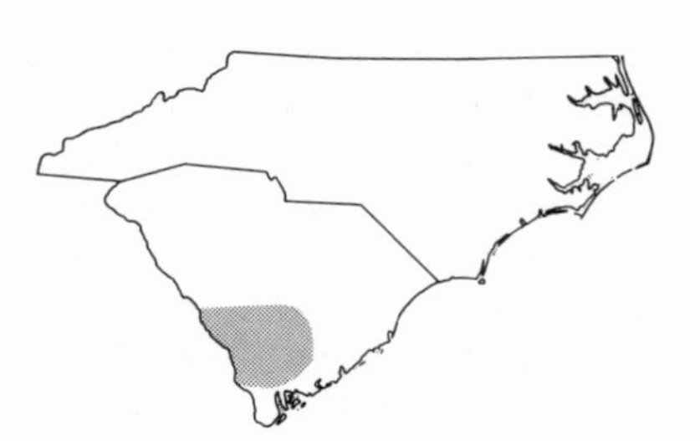
Pugnose minnow

bercles then develop around the mouth of the breeding male, as do also the bright body and fin colors that have already been mentioned. The excited male rapidly raises and lowers the dorsal fin, the bright white interior of which then resembles a flickering light. The male strongly defends a territory, usually the underside of a large, flat rock. He swims almost continuously in figure eights while under the rock, and frequently touches it with his snout or nape, appearing to clean it. He displays to the female with fins erect and attempts to lead her back to the rock. Spawning lasts but a second and occurs with the pair aligned laterally and inverted; it is repeated several times. The eggs are laid singly or in strings of two to five, and they form a single layer on the underside of the rock. The female can lay 30 to 120 eggs per spawning session and can spawn every six to seven days. One female may add eggs to the clutch of another. The female then leaves the nest, while the male remains to defend it. The eggs hatch in a few days, and growth of the larvae is rapid. The pugnose minnow no doubt reaches reproductive age in the year after it hatches. In Mississippi it is sometimes common locally and is used as a bait minnow.

Fatlips minnow

Phenacobius crassilabrum Pl. 64

Kanawha minnow

Phenacobius teretulus Pl. 66

Description. 2.4 to 4.2 inches (60 to 108 mm). These fishes have a long, cylindrical body and a round snout. The mouth is inferior, and the large fleshy lips have many papillae. Four species of *Phenacobius* occur in the mid-Atlantic region. The fatlips minnow has a light green streak above a dark lateral stripe and has two yellow spots at the base of the caudal fin; the background color is dark olive above and silvery below. The Kanawha minnow is gray-brown to olive above with small scattered black blotches, and it lacks a green streak above a dusky lateral stripe. The pelvic fin in the fatlips minnow, when pressed against the body, extends back beyond the anus while in the Kanawha minnow it does not reach it. The **suckermouth minnow**, *P. mirabilis* (Pl. 65), has a bicolored body and a dark black spot on the base of the caudal fin. The **stargazing minnow**, *P. uranops*, has a highly elongate body and eyes that are high on the head and directed upward.

Distribution and Abundance. The fatlips minnow is found only in the mountains of the upper Tennessee River drainage from Virginia to Georgia. The Kanawha minnow is restricted to the New River drainage in North Carolina, Virginia, and West Virginia. In the mid-Atlantic region the suckermouth minnow is found only in extreme western Virginia; it also occurs widely in the Mississippi River basin. The stargazing minnow occurs from western Virginia to central Tennessee and Kentucky. The fatlips minnow is encountered more commonly in North Carolina than the Kanawha minnow, while the opposite is true for Virginia. Both are uncommon, and the Kanawha minnow is listed as of spe-

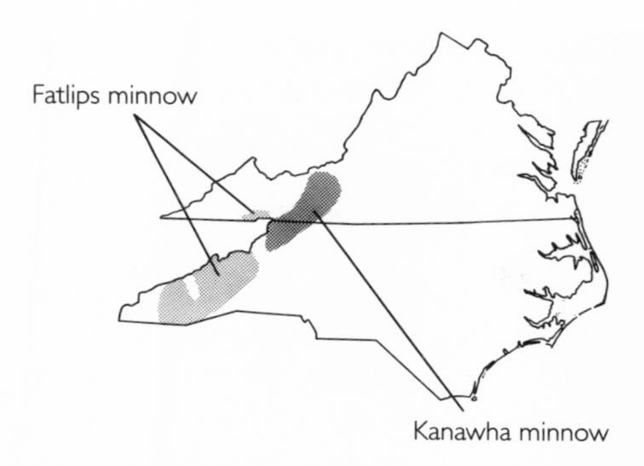

cial concern in North Carolina. The suckermouth and the stargazing minnows are common.

Habitat. These species occur in runs and riffles over gravel, rubble, and boulders in creeks to medium-sized rivers with clear water that is cool to warm. The suckermouth minnow is more tolerant of turbid conditions than are the others.

Natural History. Little is known about these fishes. The suckermouth minnow spawns from April through August, probably in gravel riffles. Apparently the other species also spawn in the spring and early summer. All four species probably feed on immature aquatic insects, worms, snails, and some plant material. The suckermouth minnow obtains food by rooting in gravel with its sensitive snout and lips.

Mountain redbelly dace

Phoxinus oreas Pls. 67–68

Description. 1.7 to 3.1 inches (44 to 79 mm). The mountain redbelly dace has scales so small that they appear to be absent. Key characteristics are a black lateral stripe that is broken underneath the dorsal fin and large black spots on the back and upper side. The breeding male is spectacularly colored, with bright red coloration on the belly, lower head, and base of the dorsal fin, black on the chin and breast, bright silver coloration on the bases of the paired fins, and fins that are mostly yellow. The similar **Tennessee dace**, *P. tennesseensis* (Pl. 69), has much smaller spots on the upper body, and it is found only in the upper Tennessee River drainage in Virginia and Tennessee.

Distribution and Abundance. The mountain redbelly dace is found in the mountains and piedmont in the central Atlantic slope from the Shenandoah River system in Virginia south to the Neuse River drainage in North Carolina and in the upper New River and Holston River drainages in Virginia. It is generally common.

Habitat. This species occurs in small headwater creeks to small rivers, usually in pools and runs over a sand, gravel, or rock bottom.

Natural History. It spawns in the spring and early summer over the nests of *Nocomis* chubs and the creek chub. It matures in one to two years and probably reaches a maximum age of three years. Its diet consists primarily of detritus and substrate slime, the latter including algae and small invertebrates.

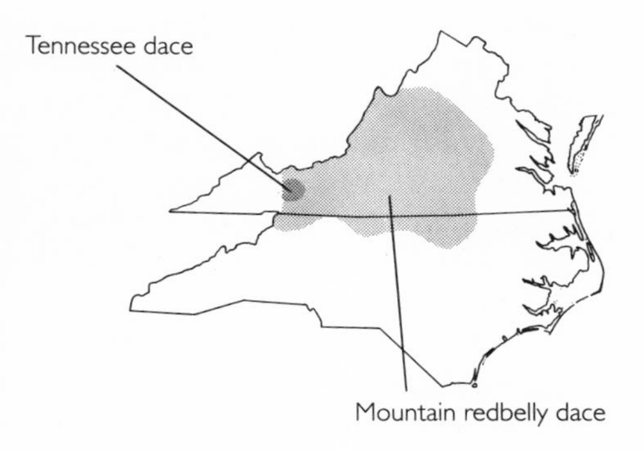

Bluntnose minnow

Pimephales notatus Pl. 70

Description. 1.6 to 4.4 inches (40 to 112 mm). The bluntnose minnow is slender, with a body approximately square in cross section, a broadly rounded snout that overhangs a small and subterminal mouth, and fins that are short and rounded. The scales on the nape are much smaller than those elsewhere on the body. The second ray of the dorsal fin is short and thickened. The back is olive yellow, with the scales distinctly dark-edged, and the side is silvery, with a dusky to black stripe. There is a conspicuous black spot at the base of the caudal fin and a black blotch in the dorsal fin. The breeding male becomes nearly black, but he has a silver bar behind the opercle, and he develops a large, spongy pad on the dorsum in front of the dorsal fin. The **fathead minnow**, *P. promelas* (Pl. 71), has a deep and compressed body with herringbone lines on the upper side. The **bullhead minnow**, *P. vigilax*, is similar to the bluntnose minnow, but it has large eyes that are directed upward.

Distribution and Abundance. The bluntnose minnow is found in the mid-Atlantic region in the upper Tennessee River drainage in the mountains of North Carolina and Virginia and on the Atlantic slope (piedmont) of Maryland and south to the Roanoke River in Virginia. It is widely distributed throughout central North America from Canada to Louisiana and on the Atlantic slope from Canada to Virginia. It is abundant, and some ichthyologists consider it to be the most common freshwater fish in eastern North America. The other two species of *Pimephales* are distributed throughout most of the central United States; they occur at a few scattered localities in the mid-Atlantic region. The fathead minnow is widely used for bait, and its release or escape has resulted in the establishment of many introduced populations.

Habitat. The bluntnose minnow is most numerous in quiet waters of medium-sized creeks to large rivers with clear, warm waters and a rock bottom.

Natural History. The bluntnose minnow is a schooling species. It has a long spawning season that extends from May to August. The male excavates a nest under a rock or other object, and the female enters the nest and attaches her eggs to the underside of the overlying object. A female can lay from 40 to 400 eggs at one time and can produce more than 2,000 eggs in one breeding season. Several females may deposit eggs in one nest, which can contain up to 5,000 eggs in different stages of development. The male vigorously guards the nest and drives away all intruders, and he cleans off sediment and dead eggs. The female matures at an age of just over one year, while the male generally matures at age two. The bluntnose minnow is primarily a bottom feeder on filamentous al-

Bluntnose minnow

gae, diatoms, small crustaceans, and larvae of aquatic insects, although it will take plankton as well as terrestrial insects at the surface. This minnow is easily propagated and is often raised and sold as a bait fish. The fathead minnow is regularly used in laboratory bioassay tests to determine the toxicity to organisms of chemicals in the water.

Sailfin shiner

Pteronotropis hypselopterus Pl. 72

Description. 1.5 to 2.5 inches (38 to 65 mm). This colorful shiner is readily distinguished from others by a broad and bluish-black stripe on the side, a deep body, and enlarged dorsal and anal fins. It is further defined by a gold stripe over the dark lateral stripe, a pale gold lower side, a blackened dorsal fin, and yellow coloration on the other fins.

Distribution and Abundance. The sailfin shiner is known only from south central South Carolina in the mid-Atlantic region. It is locally common. Overall, it occurs on the southeastern Gulf and lower Atlantic coastal plain from Alabama to South Carolina.

Habitat. It occurs in small to medium shallow creeks with a constant flow and clear water that is often tannin stained. The substrate there is usually sand and debris with some aquatic vegetation.

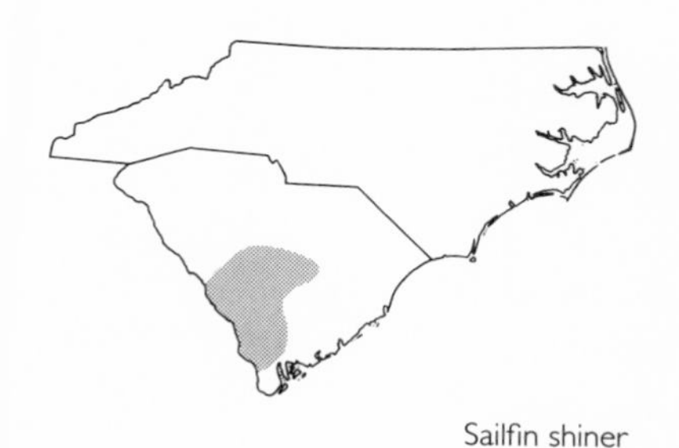

Sailfin shiner

Natural History. Nothing definitive is known about the biology of the sailfin shiner.

Blacknose dace

Rhinichthys atratulus Pl. 73

Longnose dace

Rhinichthys cataractae Pl. 74

These are small fishes with a long and slender body and a pointed snout. The scales are small, the lateral line is complete and straight, and the body is distinctly dark above and light below, usually with distinct black speckles. The genus contains nine species, of which three occur in the mid-Atlantic region.

Description. Blacknose dace: 1.7 to 3.9 inches (44 to 100 mm). A black stripe on the side from the snout through the eye and to the tail separates a brownish back from a whitish lower half. The mouth is subterminal, and there is a small barbel in the corner of the mouth.

Longnose dace: 2.9 to 6.3 inches (75 to 160 mm). This fish has a fleshy snout that extends far anterior to the inferior mouth, and a black stripe on the side that is usually indistinct, especially in larger specimens. The **Cheat minnow**, *R. bowersi*, is similar to the longnose dace, but it has larger lateral scales and a larger eye. The Cheat minnow was for long considered to be a hybrid between the longnose dace and the river chub; it is regarded as endangered in Maryland.

Distribution and Abundance. Both the blacknose and longnose daces have a large distribution in Canada and the northern United States and are at the southern extreme of their ranges in the mid-Atlantic region, where they are found in the western Carolinas, the western and northern two-thirds of Virginia, extreme northern Delaware, and most of Maryland (the longnose dace is absent from its Eastern Shore and the blacknose dace from its lower Eastern Shore). Both species are usually common.

Habitat. These daces are characteristic of small to medium-sized coolwater creeks with a slow to rapid current and a substrate of sand, gravel, or rock. The longnose dace occurs in high-gradient waters and has been reported in current speeds of over two feet per second. The blacknose dace generally occurs in slower waters. In the north, where temperatures are much lower, both species also occur in pools, lakes, and other still waters.

Natural History. Both species probably spawn from April through June in the mid-Atlantic region. Both spawn over gravel. The male blacknose dace is territorial. He spawns with several females over a period of several days. A small longnose dace male guards a territory four to eight inches in diameter. A female blacknose dace contains an average of 746 eggs, and a female longnose dace from 160 to 680, although in some areas the latter has been noted to deposit from 200 to 1,200 eggs. The eggs of the longnose dace are adhesive, transparent, and almost colorless. The eggs hatch in several days and the young grow rapidly. A young blacknose dace grows to a length of over one inch in the first fall after it hatches, and it is probably sexually mature when two years old. The longnose dace attains an age of at least five years. Both species feed on small aquatic insects, crustaceans, annelid worms, and algae.

Blacknose dace

Longnose dace

Creek chub
Semotilus atromaculatus Pl. 75
Fallfish
Semotilus corporalis Pl. 76
Sandhills chub
Semotilus lumbee Pl. 77

This genus of large (by North American standards) minnows is characterized by a thick to slightly compressed body, a broad and chunky head, a complete lateral line, a dorsal fin origin that is slightly behind the pelvic fin origin, and a small flaplike barbel located in a groove above the upper lip near, but not in, the corner of the mouth. This barbel is usually difficult for the uninitiated to see, and even for the experienced some magnifying aid is helpful.

Description. Creek chub: 3.0 to 12.0 inches (76 to 305 mm). This is a robust fish with a large black spot at the origin of the dorsal fin, a black spot, which is indistinct in large specimens, on the side just before the caudal fin, and a large terminal mouth that reaches back beyond the front of the eye. It has an olive brown back and upper side, a green silver midside with a black stripe (which is conspicuous in the young and usually weak in the adult) that extends from the snout to the tail fin, and a whitish underside. There are eight rays in the dorsal fin.

Sandhills chub: 3.1 to 9.4 inches (80 to 240 mm). This fish is similar to the creek chub, but it is usually smaller, it has nine dorsal fin rays, it lacks a black spot at the dorsal fin base (although a weak smudge may be present), and it has a much more restricted range.

Fallfish: 3.9 to 20.1 inches (98 to 510 mm). The body in this species is moderately compressed, silvery (but often with a distinct purple sheen in the spawning male), and with large scales. Large juveniles and adults have a dark smudge on the dorsal fin near its base. This is the largest native minnow in eastern North America.

Distribution and Abundance. The creek chub is broadly distributed in the mid-Atlantic region except on the Delmarva Peninsula and near the coast in the Carolinas, although there is a record of it from southern South Carolina. It also occurs in much of the eastern and adjacent western United States and Canada. The sandhills chub is found only in several counties in the sandhills area of south central North Carolina and in adjacent north central South Carolina. Because of this limited distribution, it is listed as of special concern in North Carolina. The sandhills chub and the creek chub occur together only on the edges of their respective ranges, and then only rarely. The fallfish occurs in the northern two-thirds of Virginia, most of Maryland, and northern Delaware. It is at the southern portion of its range in the mid-Atlantic region, and it occurs widely in the northeastern United States and in adjacent southeastern Canada. All three species are usually common to abundant.

Habitat. The creek chub occurs in creeks and small rivers, usually over rubble, gravel, sand, and boulders. The sandhills chub occurs in small headwater creeks, where it is often the only fish present, as well as in larger portions of creeks downstream, usually over gravel and/or sand. The fallfish prefers larger creeks and small to medium-sized rivers over rubble, gravel, and sand, although it is occasional in silt-bottomed pools.

Natural History. The reproductive male creek chub digs a nest pit in a stream by picking up and depositing mouthfuls of gravel at the upstream edge of the pit. He then guards the excavation and attempts to attract females. When he is successful, the pair spawns over the pit, and the eggs sink and fall into spaces between stones on the bottom. As eggs are deposited, the male covers them with gravel and excavates another pit just downstream; a highly successful male may produce a six-foot-long ridge of filled-in nests. Observers have reported spawning over a mound nest built of deposited stones. In the mid-Atlantic region the creek chub spawns in late March to early May. While he guards his territory against intrusion by other male creek chub, he does not attempt to prevent intrusion by other species, and numerous other species of fishes spawn on the clean gravel site he has excavated or built. The young creek chub reaches a length of three inches at the end of its first year of life and is probably mature at two years of age; the species reaches an age of at least seven years. The male grows more rapidly than the female and also reaches a larger size. Food is found by sight and consists of a wide range of aquatic and terrestrial invertebrates, as well as fishes and algae. The creek chub is an excellent bait fish. In some areas it is fished for as a desirable food species.

The sandhills chub produces nests like those of the creek chub. It probably reaches sexual maturity in one year. It feeds on a wide range of small aquatic invertebrates.

The fallfish is rare in waters that exceed a temperature of 82°. The male builds and defends stone mound nests that can be over five feet in diameter and almost three feet high. Nests are built in April and May in the mid-Atlantic region. Food is primarily insects and fishes. The fallfish reaches an age of at least six years. It is considered by some to be sporty on light tackle, and some consider it a good eating fish.

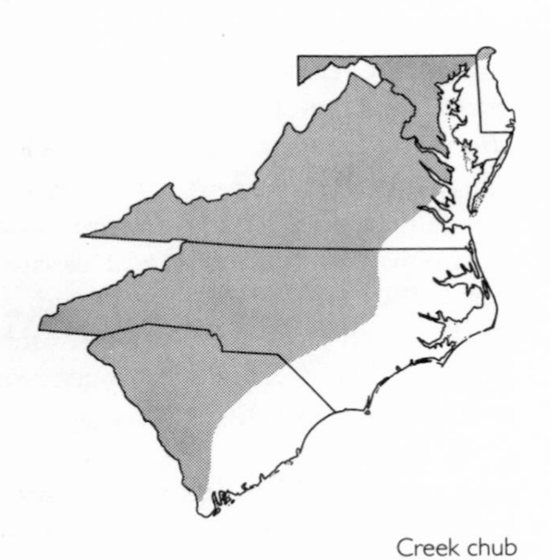

Creek chub

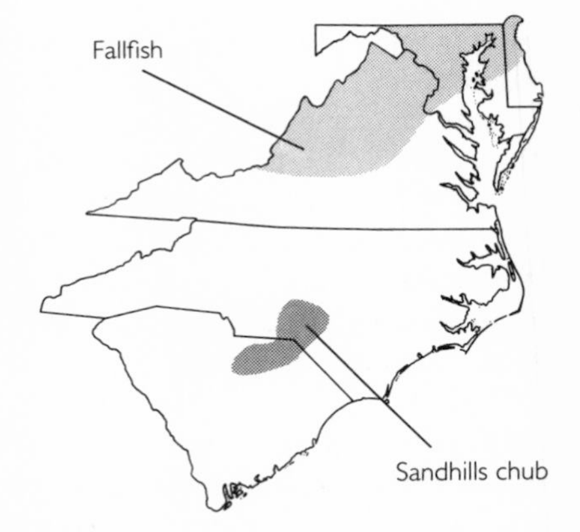

Suckers *Family Catostomidae*

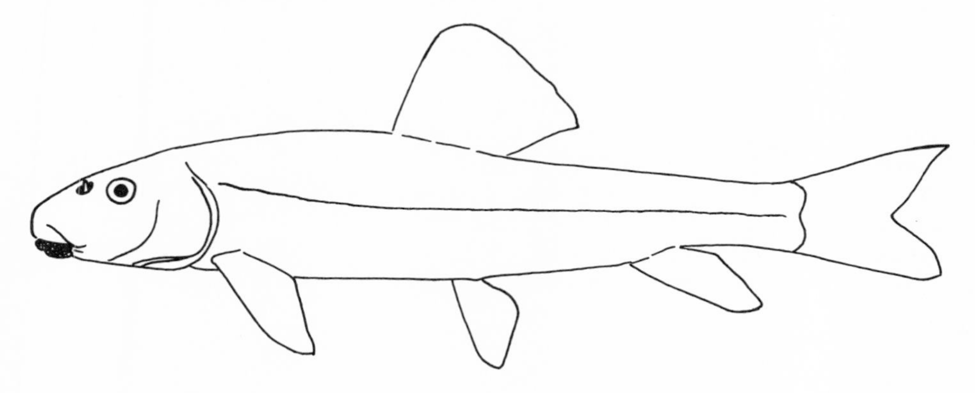

This freshwater group has representatives throughout North America, as well as in east central China and in eastern Siberia. There is a total of 12 genera with 61 species, of which 8 genera and 25 species occur in the mid-Atlantic region. The group is defined by a mouth that is usually located on the underside of the head, paired fins attached low on the body, fins supported only by rays (lacking spines), lips that are usually thick and fleshy, with their surface folded (with plicae) and/or with tiny fingerlike projections (papillae), and by one row of 16 or more teeth on each pharyngeal arch. There is only one dorsal fin, and the caudal fin is forked. The maximum length of suckers is about 40 inches, but most species are less than 24 inches in length.

Quillback
Carpiodes cyprinus
Highfin carpsucker
Carpiodes velifer Pl. 78

Description. Quillback: 12.0 to 26.0 inches (305 to 660 mm).Highfin carpsucker: 8.8 to 19.7 inches (225 to 500 mm). The quillback is a carplike, deep-bodied, laterally compressed fish with a long and falcate dorsal fin with 23 to 30 rays, a small conical head, a silver body, large conspicuous scales about twice as high as wide, and a lateral line that runs the length of the body. The lips are thick and plicate. Unlike the carp, it lacks barbels around the mouth and spines on the dorsal and anal fins. The long first dorsal fin ray usually does not reach the rear of the base of the dorsal fin, while in the similar highfin carpsucker it reaches to or beyond the base of the dorsal fin. The lips of the quillback are parallel, while in the highfin carpsucker the anterior portion of the lower lip forms a nipplelike or knoblike intrusion into the middle of the posterior edge of the upper lip. In a third species, the **river carpsucker**, *C. carpio*, there is also a nipple in the middle of the lower lip, but the dorsal fin is shorter than in the highfin carpsucker. In Missouri the quillback adult is commonly 12 to 17 inches long and weighs just less than 1 pound to 2⅓ pounds, while the highfin carpsucker adult is usually from 9 to 11 inches long and weighs less than one pound.

Distribution and Abundance. The quillback occurs in middle to upper reaches of many, but not all, major rivers of the

Carolinas and Virginia and along the upper half of Chesapeake Bay in Maryland. It is rare to common. The highfin carpsucker in the mid-Atlantic region is restricted to portions of the Cape Fear and Catawba river drainages of North Carolina, in which state it is listed as of special concern. Both species are widely distributed in the eastern and central United States. The river carpsucker is known in the mid-Atlantic region from only one site in western North Carolina near the Tennessee border, and it is listed as of special concern in that state.

Habitat. The quillback occurs in rivers and large creeks, as well as in smaller creeks if large and permanent pools are present. Outside the mid-Atlantic region, it also occurs in clear lakes. The highfin carpsucker prefers larger, deeper, and cleaner bodies of water and a firmer substrate than does the quillback. In other states, as in North Carolina, populations of the highfin carpsucker have declined drastically as a result of excessive siltation and other water quality deterioration.

Natural History. Little has been published about the quillback and its relatives in the region, but they have been studied in other portions of the United States. The quillback feeds on insect larvae and other organisms found in bottom sediments and perhaps also on the energy-rich sediments and detritus. It spawns in spring, probably from mid-April to June, and the eggs are broadcast over the bottom at the lower ends of deep gravel runs. It attains an age of at least ten years. The highfin carpsucker, on the other hand, is a summer spawner, and it also spawns over deep gravel runs. It is known to attain an age of at least eight years. The young of these two species are highly similar in appearance.

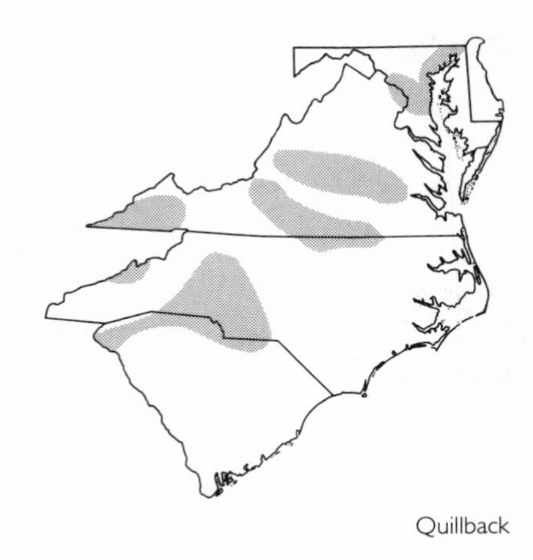

Quillback

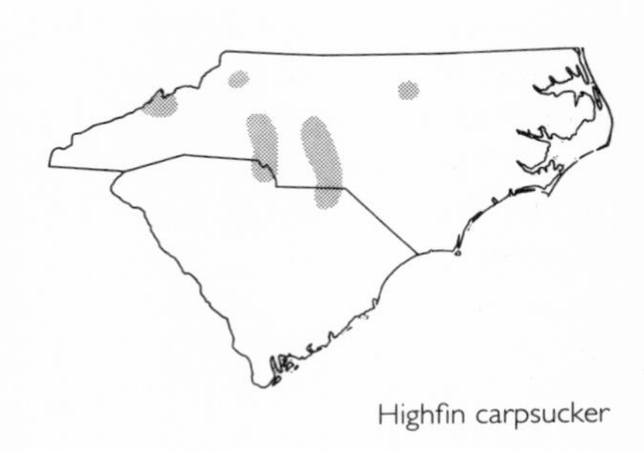

Highfin carpsucker

White sucker

Catostomus commersoni Pl. 79

Description. 10.0 to 25.0 inches (254 to 635 mm). The white sucker is an elongate, plain-patterned, whitish-gray fish, with a horizontal mouth with thick fleshy and papillose lips on the underside of the head. The scales are small and decrease in size toward the head. The lateral line runs the length of the body. The breeding male is reported to be nearly black above and white below. The young often has a series of dark blotches on the side. The closely related **longnose sucker**, *C. catostomus*, has a longer snout.

Distribution and Abundance. The white sucker is found in northern Delaware, throughout mainland Maryland, throughout Virginia except for the coastal plain, and in most of the Carolinas except the coastal area of South Carolina and the eastern half of North Carolina. This is a northern fish that is found throughout much of Canada and the United States and is at the southeastern portion of its range in the mid-Atlantic region. It is often common. The longnose sucker is known in the region only from western Maryland, in which state it is listed as endangered.

Habitat. The white sucker occurs in a wide range of habitats such as creeks, rivers, and lakes and where the water is sometimes brown (tannin stained). Where it occurs, the substrate is often gravel or rock, there is usually a pronounced current, and aquatic vegetation is usually sparse or absent.

Natural History. The white sucker is primarily a loosely schooling bottom fish. It is found in cooler waters of about 66° to 70° in the summer months. It spawns from late March through April, after it migrates to an area of gravel in creeks or small rivers at the lower ends of pools, to portions where the current speed begins to increase. While it sometimes migrates and spawns during the day, the bulk of such activity occurs at dusk, night, and dawn. The substrate is usually cleared of silt and debris by the reproductive milling of the numerous individuals above. The males typically remain over the spawning area, while females enter only when ready to spawn. Two to four males sandwich a ripe female between them and violent body tremors in both sexes accompanies expulsion of the eggs and milt. The ensuing clouds of silt, sand, and gravel help to bury the eggs. The usual number of eggs deposited per female is about 20,000 to 50,000. The eggs hatch in about 8 to 11 days at a temperature of 50° to 59°. The young for the first 10 to 11 days has a terminal mouth and feeds on microscopic animal life near the water surface. It then quickly develops the subterminal mouth characteristic of the adult and then grazes on weeds or plants that grow on solid surfaces. The near-adult and adult feed on a wider range of foods than does the young, and their mainstay is aquatic insect larvae, crustaceans, small clams, snails, and detritus. The species can reach an age of 15 years.

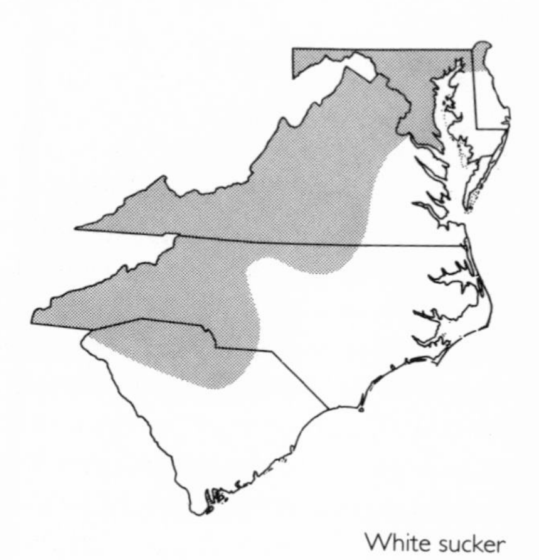
White sucker

Creek chubsucker
Erimyzon oblongus Pl. 80

Lake chubsucker
Erimyzon sucetta

Description. Creek chubsucker: 4.5 to 14.2 inches (114 to 360 mm). This sucker has a chubby body, a small, slightly oblique and nearly terminal mouth, thick fleshy lips, and prominent rounded dorsal and anal fins. There are usually 40 to 45 scales along the side of the body. There is no lateral line. The young has a whitish body with a distinct black stripe on the side that extends from the snout to the base of the tail, the subadult is light tan and has five to eight prominent black blotches on the side, and the adult is a uniform medium brown or brown with only vague remnants of the lateral blotches.

Lake chubsucker: 5.0 to 16.1 inches (127 to 410 mm). This fish appears similar to the creek chubsucker, but it has a slightly deeper body and only 34 to 39 lateral scales.

Distribution and Abundance. The creek chubsucker occurs in much of the mid-Atlantic region, although there are only a few records from the mountainous areas. The lake chubsucker in the region is known from extreme southeastern Virginia, the coastal plain of North Carolina, and the eastern two-thirds of South Carolina. Both species are widely distributed in the United States from the Gulf coast to the Great Lakes. Both are common in appropriate habitat.

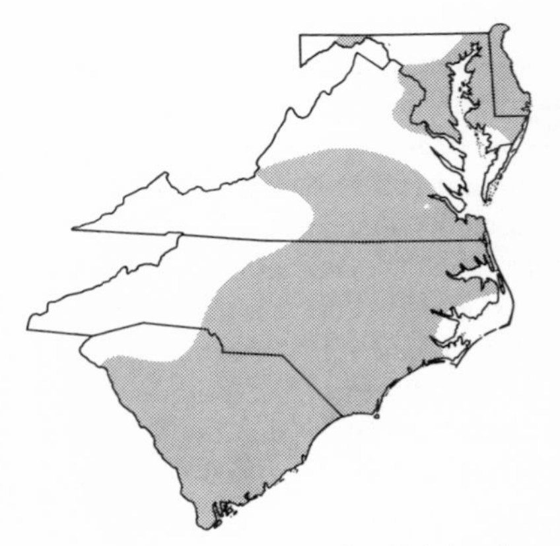
Creek chubsucker

Lake chubsucker

Habitat. The creek chubsucker occurs primarily in standing to slow-moving and clear waters of creeks, small rivers, and lakes, usually over a soft bottom with abundant aquatic vegetation, but it is not restricted to such habitats. The lake chubsucker occurs in lakes, ponds, sloughs, and impoundments and over a bottom of sand, silt, or debris; it is rare in sluggish streams.

Natural History. The creek chubsucker in North Carolina spawns from mid-March to near the end of April, when the adults migrate to small pools with a sand or gravel substrate and a current. At this time the male develops three large breeding tubercles on each side of the snout, which are used to court the female. After the several-weeks-long breeding season these tubercles drop off, leaving prominent scars. The creek chubsucker probably spawns in groups, as described for the white sucker. The number of eggs deposited per female ranges from about 9,000 to 72,000, and the larger the female, the larger the number of eggs. The young attains a length of two inches at the end of the first year and twice that by the end of the second year. The creek chubsucker feeds on small crustaceans, small insects, and algae. It can reach an age of at least six years.

The lake chubsucker spawns mostly in March and April. One ichthyologist reports that the eggs are scattered over submergent vegetation, while another reports that the male cleans an area within gravel for a nest. Eggs number

3,000 to 20,000 per female, and they are nonadhesive. They hatch in six to seven days at a water temperature of 72° to 85°, and the young then is a quarter of an inch long. The young often travel in schools with juvenile largemouth bass, adults of certain shiners, *Notropis* spp., and other fishes, all of which are light-colored with a dark lateral stripe and all of which presumably derive some benefit from schooling together. The lake chubsucker has been reported to feed on water fleas and other small crustaceans and on immature aquatic insects on or near the bottom. An age as high as eight years has been estimated for the lake chubsucker.

Northern hog sucker
Hypentelium nigricans Pl. 81
Roanoke hog sucker
Hypentelium roanokense Pl. 82

Description. Northern hog sucker: 4.0 to 24.0 inches (100 to 610 mm). This is a robust fish with a large rectangular head that is concave between the eyes. The body is cylindrical. There are three to six prominent dark saddles across the back and on the side of the brown body, the first of which is located immediately behind the head. The mouth is horizontal, and the lips are large and fleshy. The number of rays on both pectoral fins combined is 32 to 38. The female on average is longer than the male.

Roanoke hog sucker: 2.7 to 6.3 inches (68 to 160 mm). This fish is similar to the northern hog sucker, but the adult is usually much smaller and usually has only four dark saddles, the first of which is typically not located behind the head (and is only vaguely defined when it is so located). The combined pectoral fin rays number 28 to 32.

Distribution and Abundance. In the mid-Atlantic region, the northern hog sucker is known from the mountains of the Carolinas and the piedmont and the upper coastal plain in central North Carolina, most of Virginia except for the coast, Maryland west of Chesapeake Bay, and the piedmont in Delaware; it is absent from most of the Delmarva Peninsula. This fish is widely distributed in the eastern United States and is usually common. The Roanoke hog sucker occurs only in the mountains and the central piedmont, where it is limited to portions of the Roanoke River drainage in south central Virginia and adjacent north central North Carolina. It is uncommon.

Northern hog sucker

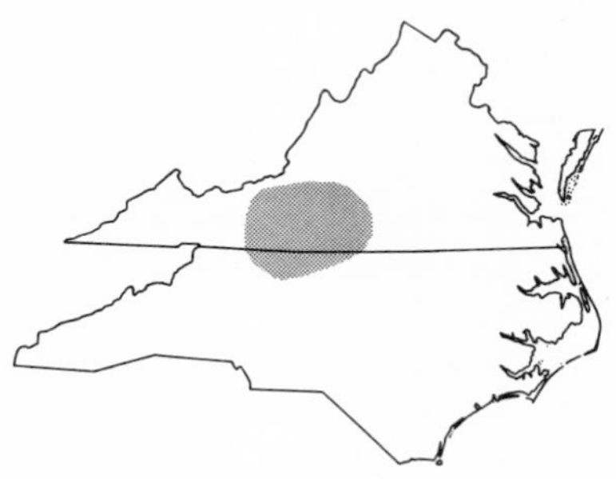
Roanoke hog sucker

Habitat. The northern hog sucker in the warmer months is usually found in clear and shallow high-gradient creeks and small rivers, where it prefers riffles and pools with a gravel and rubble substrate. It also occurs over a substrate of boulders, exposed bedrock, and silt. This species moves downstream to deeper low-gradient waters for the winter. It remains there until the water temperature rises to above 40°, when it begins its upstream migration. The Roanoke hog sucker occurs in creeks to small rivers in riffles and runs over gravel, rubble, or boulders. It is only occasional in areas of slower current and over a substrate of finer materials.

Natural History. The northern hog sucker spawns from late March to June and begins to spawn when the water temperature reaches 60°. The male moves to riffles three to five inches deep over areas of gravel, and the female joins him from nearby pools. Two or more males closely crowd a female, and the eggs and sperm are emitted by the tremoring and shuffling fishes. The actual emission of gametes lasts for two seconds and is repeated every few minutes. The eggs are broadcast, sink, remain where deposited in spaces between gravel, and are abandoned by the parents. The young grows at a rate of approximately two inches per year. Most males mature in two years and some in one, and most females mature in three years. The species reaches an age of at least 11 years. The food is insect larvae, crustaceans, and low encrusting vegetation on rocks. Bracing itself on its extended pectoral and pelvic fins, with its back slightly arched, it roots for food in substrate sand and detritus with its protrusible jaws and rapidly voids useless sand and debris from its gill openings, working like an efficient mining machine. It can excavate rapidly, while the eyes placed high on its head still allow it to watch out for predators; it then quickly moves off to repeat the digging process. It also sucks and scrapes food off rocks and turns rocks to explore beneath them. While edible, the flesh of this fish is not favored.

The Roanoke hog sucker breeds in mid- and/or late spring. The male may mature in one year but usually requires two, and the female matures in three. The male is known to live up to four years, the female to five. Its food is presumably similar to that of the northern hog sucker.

Smallmouth buffalo

Ictiobus bubalus Pl. 83

Description. 15.0 to 30.7 inches (381 to 780 mm). The smallmouth buffalo is a carplike fish; but the body is deeper than that of the common carp, and the head is smaller and more conical. There are distinct grooves on the thick upper lip. The mouth is small and subterminal, and barbels are absent. The adult has a strongly keeled nape; it is olive or bronze on the back, black to olive yellow on the side, and pale below. The fins are olive to black, often bluish, and give rise to a local name of "bluefin sucker." The edge of the subopercle (the last and lowest of the four bones that comprise the operculum) is approximately semicircular in shape on the outer edge. The **bigmouth buffalo**, *I. cyprinellus*, is similar, but it has a terminal and sharply oblique mouth.

Distribution and Abundance. In the mid-Atlantic region the smallmouth buffalo is found only in a few localities in western North Carolina and central North and South Carolina; it is probably native only to western North Carolina. It also occurs in much of the eastern United States from the Great Lakes to the Gulf of Mexico, primarily in the Mississippi River basin, and as far west as the upper Missouri and upper Rio Grande rivers. It is locally common and occasionally supports commercial fisheries, as in the Pee Dee River in North Carolina. The bigmouth buffalo occurs in a few localities in south central North Carolina.

Habitat. The smallmouth buffalo is a species primarily of large rivers, but it is occasional in medium-sized rivers and lakes. It prefers pools, oxbow lakes, and deeper waters, although it often occurs in a pronounced current. It prefers clear, clean water.

Natural History. This fish is primarily a bottom feeder on attached algae, insect larvae, clams, detritus, and zooplankton. Like many fishes, it is no doubt opportunistic and feeds on those foods most available. It spawns from May to September, at a water temperature of 66° to 82°. The males concentrate on shoals in water four to ten feet deep in March and April, and additional males and females join as they ripen and leave as they are spent. Tens of thousands to hundreds of thousands of eggs per female are broadcast over shoals and sink to a variety of substrates below. The young tend to school. The male matures earlier than the female, which in New Mexico has been reported as early as at ages of one and two years, respectively, and in colder South Dakota as late as at ages of 9 and 11 years, respectively. It is caught commercially in some areas, and it is a tasty food fish. Competition with the introduced carp has reportedly reduced its numbers in some areas.

Spotted sucker

Minytrema melanops Pl. 84

Description. 5.9 to 19.5 inches (150 to 495 mm). The spotted sucker resembles the white sucker in color and body shape, but it differs from the latter with its 8 to 12 parallel rows of dark spots that run the length of the body and a dusky to black edge on the dorsal fin and on the lower lobe of the caudal fin. It has a small horizontal mouth with thin fleshy lips, and no, or an incomplete, lateral line.

Distribution and Abundance. This sucker has been recorded from the southeastern quarter of North Carolina and from widely scattered localities throughout most of South Carolina. It is found in much of the eastern United States from the Gulf coast to the Great Lakes, and in the mid-Atlantic region it is at the eastern edge of its range. It is often moderately common but has suffered from the effects of siltation and other pollutants.

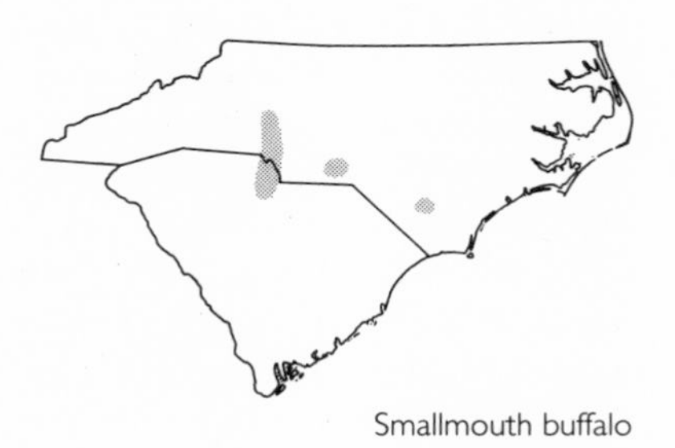

Smallmouth buffalo

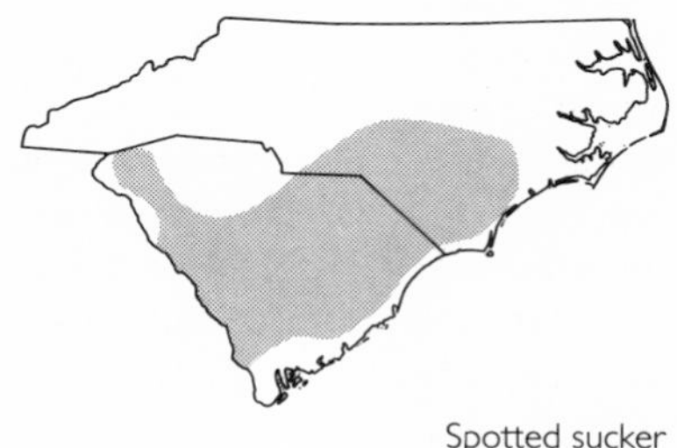

Spotted sucker

Habitat. The spotted sucker prefers large pools of small to medium-sized clear rivers over a firm substrate of clean sand, gravel, or hard clay, but it is occasional in turbid waters and in creeks, lakes, sloughs, oxbows, large rivers, and impoundments.

Natural History. This species spawns from March through May at a water temperature of 53° to 67°. The eggs are deposited in riffles above large pools. Two ripe males then usually sandwich one female between them as they settle on the bottom and face upstream. The males clasp the rear of the body of the female and vibrate vigorously, during which the eggs are shed and fertilized. At the finish of spawning, the three often break the surface of the water as they separate. The eggs hatch in 7 to 12 days. This sucker attains sexual maturity at an age of three years, and the maximum recorded age apparently is six years. Food is mostly crustaceans and insect larvae.

Black redhorse
Moxostoma duquesnei Pl. 85
Shorthead redhorse
Moxostoma macrolepidotum Pl. 87
V-lip redhorse
Moxostoma pappillosum

The genus *Moxostoma* is widely distributed from central Canada to central Mexico. The genus is distinguished by 12 rows of scales around the caudal peduncle, thick fleshy lips, a body that is robust to long and slender, a large horizontal mouth, and a complete lateral line. Some species have a large range, while several are extremely localized. These fishes occur throughout much of the mid-Atlantic region but not on most of the Delmarva Peninsula. Only one species or a very small number of species is likely to occur in any given area. These are fishes of mostly large creeks and rivers, and most are found in faster waters of the mountains and the piedmont, although some occur in lakes and impoundments. Seven species of *Moxostoma* occur in the region.

Description. Black redhorse: 8.0 to 20.0 inches (204 to 510 mm). The body is long and slender, as is the caudal peduncle. The lower lip is plicate and has a broadly V-shaped rear edge. The pectoral, pelvic, and anal fins are light orange, and the other fins are slate gray. The dorsal fin edge is usually concave, there are 45 to 47 lateral line scales, and there are 12 to 14 rays on the dorsal fin as well as 9 or 10 rays on the pelvic fin. The **golden redhorse**, *M. erythrurum* (Pl. 86), is similar but has 40 to 42 lateral line scales, a shorter and deeper caudal peduncle, and only nine pelvic fin rays.

Shorthead redhorse: 10.1 to 18.3 inches (258 to 464 mm). This species has a stout body and a short head, which comprises about 20 percent of the total length. The plicae of the lower lip are transversely subdivided, and the lower lip has a nearly straight rear edge. The pharyngeal teeth are comb tooth–like. The caudal and dorsal fins are red or slate-colored and the other fins range from cream to yellow to red. The dorsal fin is concave and has 12 or 13 rays. The **river redhorse**, *M. carinatum,* is similar but has a larger head, thicker lips, a slightly V-shaped rear edge on the lower lip, and molarlike pharyngeal teeth. The **robust redhorse**, *M. robustum,* is similar to the river redhorse, but it has ten pelvic rays instead of nine. This last species is not the same as the one called smallfin redhorse, *M. robustum,* by recent authors. Edward D. Cope discovered and described the robust redhorse from the Pee Dee drainage in North Carolina in 1870. Unfortunately, later ichthyologists misinterpreted his specimens and placed the name of *robustum* on the fish that became known as the smallfin redhorse (see the *Scartomyzon* account below). During the period from 1980 to 1992, specimens of a large redhorse were collected in the Savannah River below Augusta, Georgia (two fish), the Pee Dee River in North Carolina (one fish), and in the Oconee River in Georgia (many). These fish were determined to be the true robust redhorse as described by Cope in 1870 and had been unrecognized as such for 110 years.

V-lip redhorse: 9.9 to 16.5 inches (252 to 420 mm). The body is long and slender, and the rear edge of the lower lip is acutely V-shaped; both lips are covered with papillae. The dorsal fin is concave and has 12 or 13 rays. The **silver redhorse**, *M. anisurum*, is similar, but it has a stouter body, and a straight or convex dorsal fin with 14 to 16 rays.

Distribution and Abundance. Most of these species occur throughout the region as far north as central Virginia, although the shorthead redhorse occurs north to Maryland. They are generally uncommon, but some species are locally abundant. The river redhorse is considered to be of special concern in North Carolina. The robust redhorse is a candidate for listing as endangered by the federal government.

Habitat. Moxostoma occurs in medium to large creeks and rivers, in cold to warm and in high- to low-gradient waters, and usually over gravel, rock, or boulders but sometimes also over silt. Some occur in natural lakes and in impoundments.

Natural History. Their food is mostly small crustaceans, insects, mollusks, al-

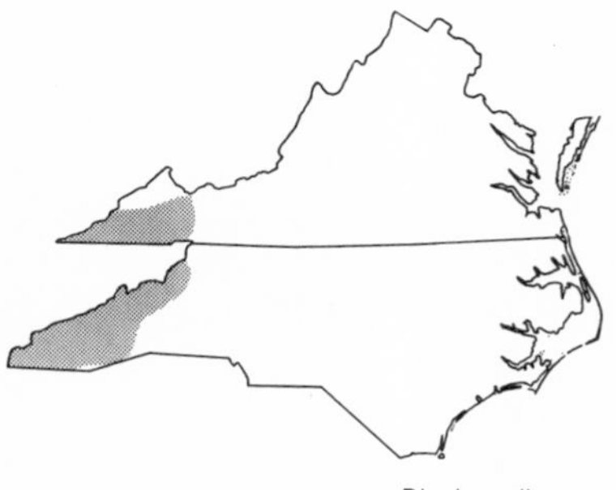

Black redhorse

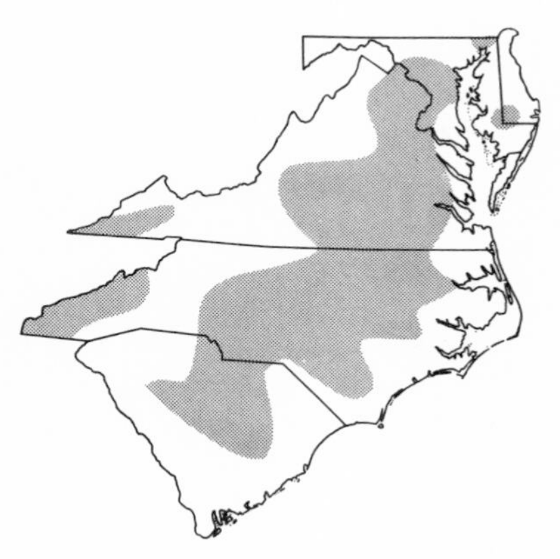

Shorthead redhorse

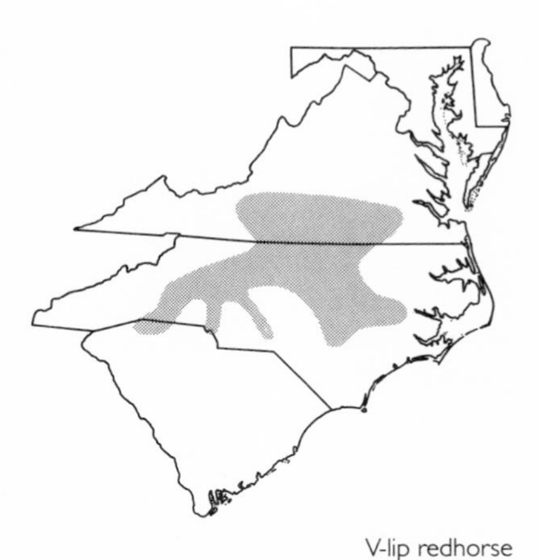

V-lip redhorse

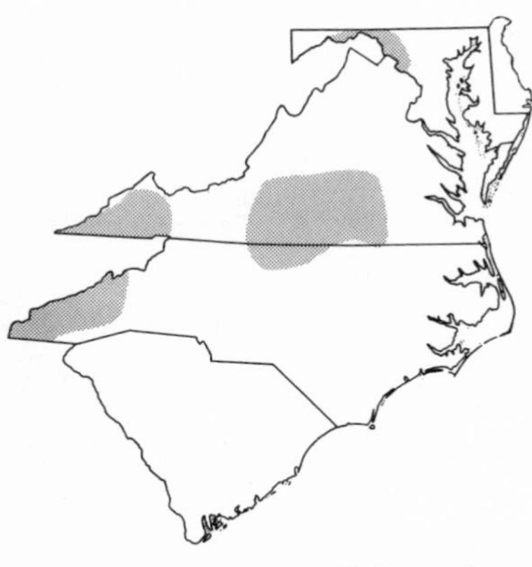

Golden redhorse

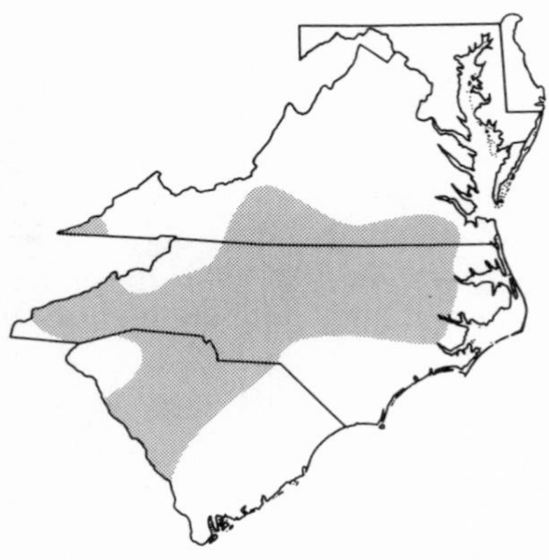

Silver redhorse

gae, and detritus. They spawn from March to May, when some species gather in large groups over shallow gravel riffles. Thousands of eggs may be produced per female. Growth is rapid, especially in the second year, and redhorses reach sexual maturity in four or five years. Several species can attain an age of at least nine years, and the robust redhorse can live up to 20 years. Some of the larger species are sought by commercial and other fishermen for food and sport. Because of the larger, deeper, and faster bodies of water that they usually inhabit, specimens are difficult to catch, and thus we know relatively little about them.

Bigeye jumprock
Scartomyzon ariommus Pl. 88
Black jumprock
Scartomyzon cervinus Pl. 89
Striped jumprock
Scartomyzon rupiscartes

The jumprocks occur from the James River drainage in Virginia to the Apalachicola River drainage in Alabama. Four of the five species in the genus are found in the mid-Atlantic region. Jumprocks are similar to the redhorses, but the former have 16 rows of scales around the caudal peduncle and the latter 12 rows. Jumprocks also usually have dark and light stripes on the body.

Description. Bigeye jumprock: 5.2 to 9.0 inches (132 to 228 mm). This distinctive fish has a long cylindrical body that is flattened in front and compressed behind. The eye is very large, and the top of the head is slightly concave. The lips are flat and flaring and covered with papillae. The body color is olive to brown on the dorsum and white on the venter.

Black jumprock: 4.1 to 7.5 inches (105 to 190 mm). This is one of the more distinctive *Scartomyzon*, with a long and cylindrical body, black or dusky tips on the dorsal and caudal fins, at least seven dark lateral stripes per side that extend from the head to the tail, plicate lips, and a straight edge on the lower lip. The young has about four dark blotches on the side.

Striped jumprock: 6.1 to 11.0 inches (156 to 280 mm). The body shape is similar to that of the black jumprock, the head is slightly convex or flat between the eyes, the snout is rounded, and the upper lobe of the tail fin is pointed and the lower lobe rounded. The body is light brown, the side has numerous prominent brown stripes that are wider than the intervening white stripes, the venter is whitish, and the fins are dull orange-red. There is usually a dusky edge on the dorsal and caudal fins. The dorsal fin has 10 or 11 rays. The **brassy jumprock** (Pl. 90), a not yet described species of *Scartomyzon*, is similar, but it has weak brown stripes that are more narrow than the white stripes, usually 12 dorsal fin rays, a distinctly convex head between the eyes, and fins that lack a dusky edge. This is the species that has in error been called the smallfin redhorse, *M. robustum*, during past decades.

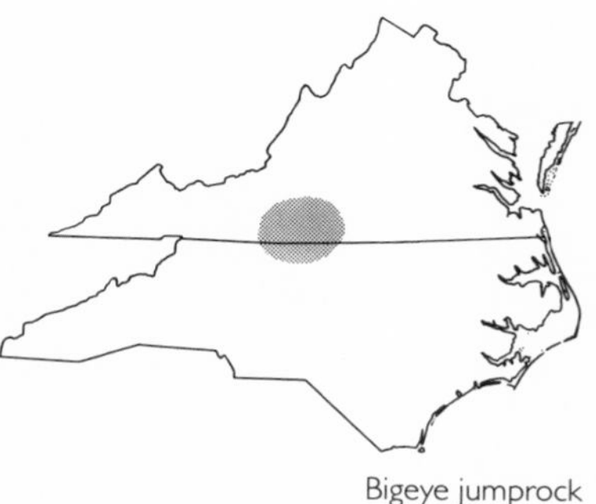

Bigeye jumprock

Distribution and Abundance. The bigeye jumprock is restricted to the upper Roanoke River drainage in Virginia and North Carolina. The black jumprock occurs from the James River drainage in Virginia to the Neuse River drainage in North Carolina. It also has been introduced into the New River drainage in Virginia. The striped jumprock is widespread from the Santee River drainage in North Carolina to the Chattahoochee River system in Georgia. It has been recently found in the upper Pee Dee River drainage, into which it might have been introduced. The brassy jumprock occurs from the upper Cape Fear River drainage in North Carolina to the Oconee River system in Georgia. The bigeye jumprock is considered to be of special concern in North Carolina. The other three species can be locally common.

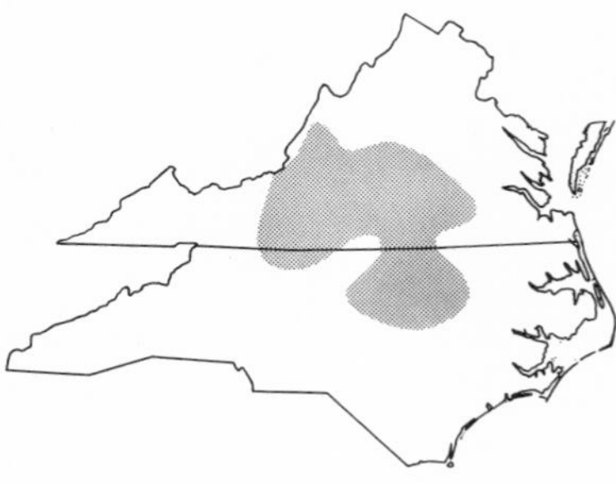

Black jumprock

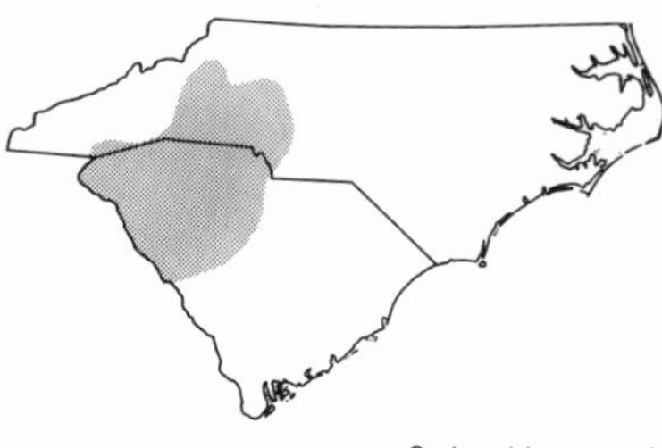

Striped jumprock

Habitat. Jumprocks usually occur in warm, small to large rivers, where they occupy runs and riffles with a rubble, gravel, and sand bottom, as well as in fast and deep water, often near bedrock and boulders.

Natural History. Jumprocks feed on aquatic insects and water mites; the black jumprock feeds on the bottom, often in groups. The group apparently spawns from March to May. Growth is rapid and sexual maturity can be reached in two years.

Rustyside sucker
Thoburnia hamiltoni Pl. 91
Torrent sucker
Thoburnia rhothoeca

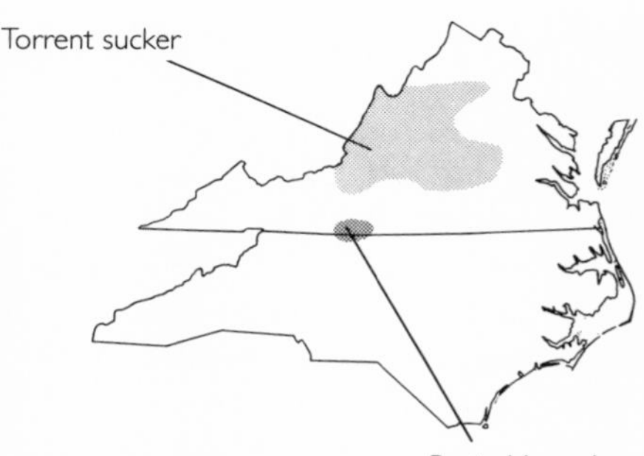

Description. Rustyside sucker: 2.7 to 7.1 inches (68 to 180 mm). This species is distinguished from other suckers by two prominent pale or dusky areas on the base of the caudal fin, highlighted by black streaks along the adjacent caudal fin rays. Each half of the lower lip is squarish or broadly rounded and has more papillae (outward-pointing projections) than plicae (forward-to-backward oriented folds of skin). There are four or five weak dark brown blotches on the side, which are sometimes connected to dark saddles on the back. The breeding male has a rusty red lateral stripe, and the fins are creamy, pale yellow, or pale olive.

Torrent sucker: 2.7 to 7.1 inches (68 to 180 mm). This fish is almost identical to the rustyside sucker, but it has a smaller lower lip whose halves are roughly triangular in shape and on which are located more plicae than papillae.

Distribution and Abundance. The rustyside sucker is restricted to the upper Dan River system in Patrick County in western Virginia and in immediately adjacent Stokes County in North Carolina. It is locally common to rare. Because of its tiny range, it is considered to be endangered in North Carolina. The torrent sucker is restricted to central Virginia and to one or two adjacent streams in West Virginia. It is usually common.

Habitat. Both species occur in clear, swift waters of creeks and small rivers in rocky riffles and runs. The young prefer the calmer waters of runs and pools.

Natural History. The rustyside sucker feeds on the larvae of aquatic insects and detritus. It probably breeds from late March to early May. It reaches sexual maturity at an age of two or three years, and it attains a maximum age of four years. The torrent sucker feeds on plants and midge larvae. It spawns in April and May. The maximum age reached is seven years.

Bullhead Catfishes *Family Ictaluridae*

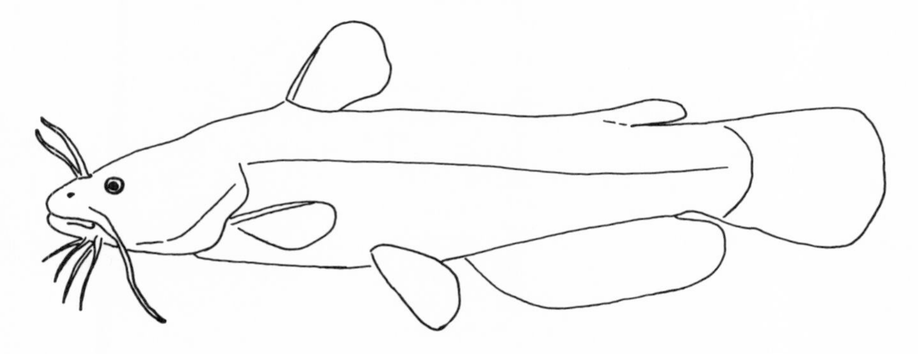

The bullhead catfishes are mostly nocturnal and omnivorous freshwater bottom feeders. They are easily identified by a lack of scales, a sharp spine in the pectoral and dorsal fins, and an adipose (fleshy) fin located on the back between the dorsal and caudal fins. Most have a small eye, a dark dorsum, and a whitish venter. Four pairs of whiskerlike barbels located near the mouth function as organs of touch and taste and aid in navigation and finding prey in turbid waters and at night. The large mouth is usually well equipped with patches of bristlelike teeth. A mild venom, which is a defense against predators, is associated with the dorsal and pectoral fin spines. It can cause a painful wound to a careless person.

This diverse group ranges from Canada to Guatemala, and contains about 45 species. Eighteen species occur in the mid-Atlantic region. These include nine species of small madtoms that never reach more than a few inches in length and are seldom seen by fishermen. Five other species are bullheads, which are medium-sized fishes that reach a weight of a pound or two and are often caught by fishermen. And there are four species of large catfishes, of which the largest, the blue and the flathead catfishes, may attain a length of nearly six feet and exceed a weight of 100 pounds. All of the large catfishes in the region are commercially important.

Snail bullhead
Ameiurus brunneus
Flat bullhead
Ameiurus platycephalus Pl. 92

Description. These two similar species have a flat head, a short and rounded anal fin, and a large dusky blotch at the base of the dorsal fin.

Snail bullhead: 6.9 to 11.4 inches (175 to 290 mm). This species has a uniformly colored maxillary (upper lip) barbel, a decurved snout in profile, a yellow-green or olive dorsum, a gold to dusky yellow side, and a blue-white to white venter.

Flat bullhead: 7.0 to 11.4 inches (179 to 290 mm). This species can be distinguished from the snail bullhead by a bicolored maxillary barbel, a relatively straight profile of the snout, and a gold-yellow to dark brown body above, dark mottling on the side, and a dull cream-colored belly.

Distribution and Abundance. The snail bullhead is found in the mountains, piedmont, and coastal plain from the

Dan River system in Virginia and North Carolina south to the Saint Johns River drainage in Florida. It is common. The flat bullhead occurs in the mountains, piedmont, and coastal plain from the Roanoke River drainage in Virginia south to the Altamaha River drainage in Georgia. It is common to uncommon.

Habitat. The snail bullhead is typical of rocky riffles, runs, and flowing pools in fast streams. The flat bullhead occurs in backwater areas of large rivers, lakes, impoundments, and ponds on a mud, sand, or rock bottom. The young of both species inhabit smaller and clearer streams.

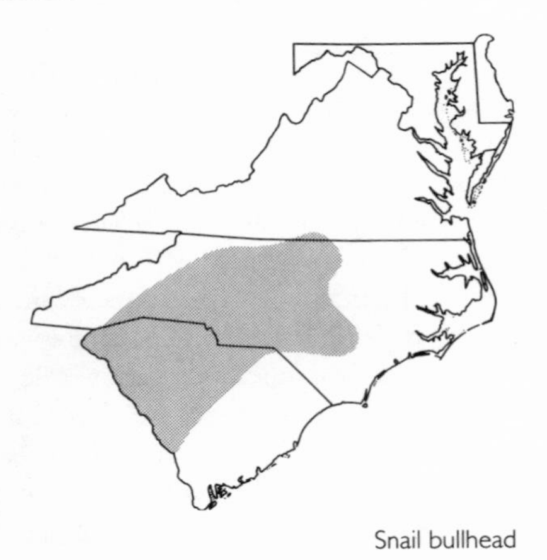

Snail bullhead

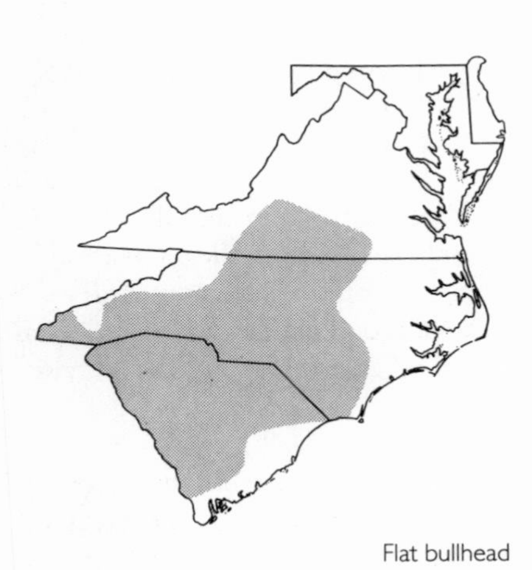

Flat bullhead

Natural History. The snail bullhead spawns from March to June in North Carolina. The flat bullhead spawns in June and July. The latter species attains maturity at an age of three years and can live up to seven years. The snail bullhead is omnivorous and feeds mainly on snails (hence its common name), filamentous algae, larvae of aquatic insects, mollusks, and fishes. The flat bullhead eats a wide range of animal food such as bryozoans (moss animals), snails, insects, and fishes.

White catfish

Ameiurus catus Pl. 93

Description. 8.2 to 24.4 inches (208 to 620 mm). The white catfish is bluish-gray above and shades to a silvery white on the belly; the side is unmarked. The pectoral and pelvic fins are uniformly light in color. The tail fin is weakly to moderately forked and the tips are rounded. There are usually 19 to 23 rays in the anal fin, and its margin is convex.

Distribution and Abundance. This species is native to the Atlantic coastal states from New York to Florida, and it occurs in most river systems. It has been widely transplanted into lakes and reservoirs and into the Tennessee River drainage in western North Carolina.

Habitat. The white catfish is found in lakes and streams, primarily in the piedmont and coastal plain. It is occasional in brackish water.

Natural History. The white catfish is the smallest of the large catfishes. Most individuals caught weigh less than three pounds, although occasional specimens weigh 10 to 15 pounds. It feeds on aquatic plants, insects, and fishes. It is an early

White catfish

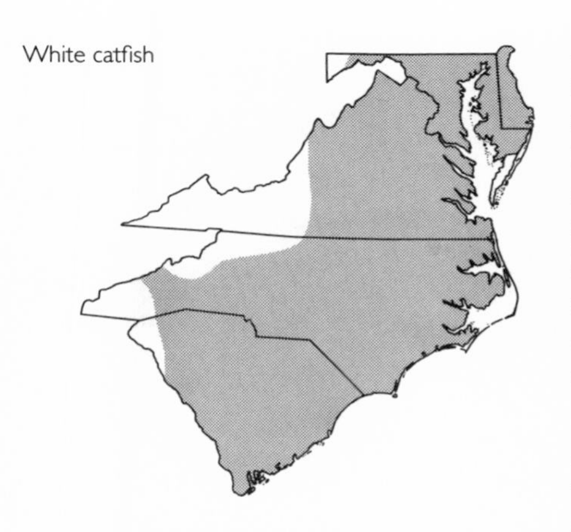

summer spawner. Both members of a mated pair build a nest by fanning the bottom, creating a depression 30 to 36 inches wide and up to 18 inches deep. Several thousand eggs are deposited there. The eggs are guarded until they hatch, usually within six to seven days. Sexual maturity appears to be reached by the third or fourth year, when the fish is about seven to nine inches in length, and the maximum life span is about 11 years. It can be caught with worms, small fishes, chicken liver, shrimp, and various commercial baits. While this species feeds mostly at dusk or at night, it can also be caught during the day. The flesh is good if the fish is taken from clean waters. It is often stocked in ponds.

Yellow bullhead

Ameiurus natalis Pl. 94

Description. 6.5 to 18.5 inches (165 to 470 mm). A distinguishing character of this catfish is its white to yellowish chin barbels. The anal fin is moderately long and has an almost straight margin, often with a dark midstripe, and 24 to 27 rays. The far edge of the tail is rounded or nearly straight. The body grades from olive to yellow on the back and to bright yellow on the side or white on the belly.

Distribution and Abundance. The native distribution of the yellow bullhead is throughout the eastern and central United States. It occurs naturally in all of the mid-Atlantic region except for a few mountainous areas of western North Carolina and Virginia and the Delmarva Peninsula. It has been widely introduced elsewhere. It is most common on the coastal plain.

Habitat. This species is found in shallow waters where vegetation is dense to absent over a soft bottom. It occurs in clear lakes, ponds, slow creeks and rivers, and impoundments. It is usually common. The yellow bullhead tolerates polluted waters better than most catfishes.

Natural History. Food is found with the barbels on the bottom at night and includes plant matter, crayfishes, aquatic insects, mollusks, and fishes. Some persons consider the yellow bullhead to be a scavenger. In the mid-Atlantic region it spawns in spring or early summer. Both sexes excavate a nest depression in shallow water among weeds, logs, roots, or other cover, or they burrow several feet into a stream bank. Several hundred glutinous eggs are released and fertilized

Yellow bullhead

at a time and then attach to logs, roots, or other material. The eggs number in the thousands and hatch in five to ten days. The larvae and juveniles are guarded by the male until they are about two inches long and capable of caring for themselves. Maturity is reached when three years old. This species is one of the best bullheads for sport, and fishing for it is best at night, dawn, or dusk, when it is most active. It can also be caught during the day. Worms, meat scraps, or cheese make effective bait. The flesh is excellent eating.

Brown bullhead

Brown bullhead

Ameiurus nebulosus Pl. 95

Description. 5.9 to 20.9 inches (150 to 532 mm). This species is yellow-brown to gray or black, and it is often mottled or spotted. It is also identified by five to eight large sawlike teeth located on the rear of the pectoral fin spine. Such teeth are tiny or lacking in the similar **black bullhead**, *A. melas.*

Distribution and Abundance. The brown bullhead is native to much of the eastern half of the United States. It is found in all of the mid-Atlantic region except western Virginia and southwestern North Carolina. It is common in the northeastern parts of its range and on the Atlantic and Gulf coastal plains. It has been widely introduced outside its native range. The black bullhead has a much smaller distribution in the mid-Atlantic region, and it is restricted to the western portions of Virginia, North Carolina, and South Carolina, where it is occasional.

Habitat. The brown bullhead is found in vegetated areas of pools, slow moving creeks, small to large rivers, impoundments, lakes, and ponds over a soft bottom.

Natural History. The brown bullhead spawns in spring and summer, from early morning to early afternoon. A pair prepares a circular nest in a substrate of sand, gravel, or mud, and under the shelter of a log, rock, or vegetation. The male drives other bullheads from the nest area. The spawning act begins with the male and female swimming in a circle, one behind the other, near the opening of the nest, continues with both resting side by side in the nest, facing in opposite directions, and culminates with the depositing and fertilizing of eggs while both remain in this position. This sequence is repeated many times until the female is spent. A nest may contain up to 13,000 eggs, which are about one-eighth inch in diameter and attached to each other to form a cluster. The parents guard the eggs until they hatch in five to ten days into one-eighth to one-half inch long larvae laden with a large yolk sac; they resemble small tadpoles. The larvae are guarded and shepherded about while kept in a tight school, or "ball," by the parents. Fry that stray are picked up in the mouth of either parent and returned to the brood. The adults stay with the brood until the young are about two inches long.

Sexual maturity is reached at an age of two to three years, and the maximum age attained is six or seven years. This species is a typical omnivorous, bottom-feeding catfish. It consumes crustaceans, insects, worms, algae, mollusks, and fishes. It feeds primarily at night or late in the afternoon.

As with other catfishes, its varied diet allows the angler a wide selection of bait. The common earthworm, crickets, minnows, liver, and commercial or homemade bait with a foul odor are effective. Use a long-shanked hook since all bullheads are noted for swallowing the bait and a long shank allows easier hook removal. The brown bullhead is eaten with gusto by many.

Blue catfish

Ictalurus furcatus Pl. 96

Description. 20.0 to 65.0 inches (508 to 1,650 mm). This large catfish is similar to the white catfish. The back is bluish and markedly humped at the dorsal fin, and the belly is silvery white. The tail is deeply forked, and there are usually 30 or more rays in the straight-edged anal fin. The pectoral and pelvic fins are usually light-colored, with a dusky margin. A new record hook-and-line catch was established in 1991 when a fish of 109 pounds, 4 ounces, was taken from the Santee-Cooper Reservoir in South Carolina. Historic records indicate that in the Mississippi River basin in the late 1800s the blue catfish reached a weight of over 300 pounds.

Distribution and Abundance. This catfish is native to the major rivers of the Mississippi River basin and to adjacent Gulf coastal drainages. In South Carolina, North Carolina, and Virginia it has been stocked in rivers, large lakes, and reservoirs. It is often common.

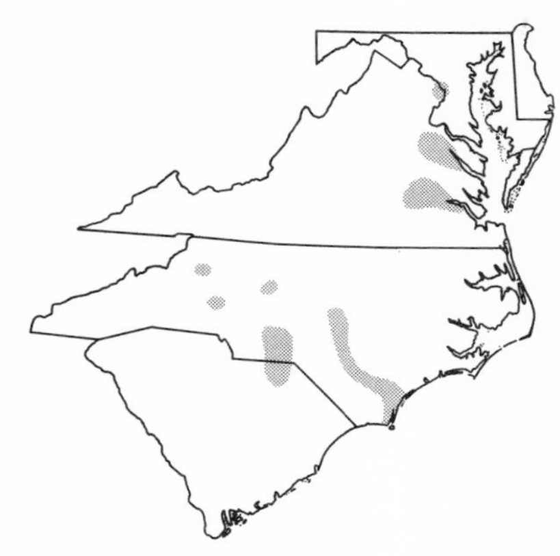

Blue catfish

Habitat. The blue catfish inhabits deep pools of large rivers. It prefers swift water over a sandy bottom, but it also does well in large impoundments.

Natural History. The food of the blue catfish includes detritus, insects, mollusks, crayfishes, and fishes. It is an early summer spawner. A nest is fanned out under a log or an undercut riverbank, and the eggs are laid in a cavity or depression and guarded by the parents until they hatch. The maximum known age is 21 years, but based on the large specimens caught by fishermen, it probably lives longer. It is a valuable game and food fish.

Channel catfish

Ictalurus punctatus Pl. 97

Description. 10.5 to 50.0 inches (267 to 1,270 mm). The channel catfish is more slender than the other large catfishes. The back is bluish gray, and it shades to tan on the side and to white on the belly. The tail fin is deeply forked. There are usually from 24 to 29 rays in the anal fin, which has a convex edge. The young has scattered small dark spots on the side, and these are absent in otherwise similar-appearing catfishes.

Distribution and Abundance. In the mid-Atlantic region this catfish is probably native only to westward-flowing river drainages, while it is widely distributed in the Mississippi River basin and its tributaries from southern Canada to the Gulf coast. It has been widely stocked in recent years, however, and it now occurs in suitable habitat throughout much of the United States. It is generally common.

Habitat. The channel catfish is native to medium-sized and large rivers. It prefers clear and clean water with some current, and it is often found over a sandy or rocky bottom. It is highly adaptable, however, and also thrives in farm ponds, lakes, and reservoirs.

Natural History. The food of this active catfish ranges from plant seeds to insects, crayfishes, and fishes. It feeds primarily on the bottom and mostly at night. This catfish is, however, more active in daylight than most catfishes. The male constructs a nest in spring or early summer. He often spawns with several females, and he then guards the eggs and fry until the young leave the nest. The female spawns but once per year, and can lay up to 21,000 eggs, which look like a mound of yellow tapioca. The channel catfish matures when four to six years old; few individuals live beyond seven years, but some attain a maximum age of at least 15 years.

The channel catfish is one of the most important freshwater fishes in the mid-Atlantic region: it is widely stocked in farm ponds for sport fishing, it is the species that is usually raised in aquaculture operations, and it is the one most often caught commercially and recreationally in the wild. Catfish meat served in restaurants and found at supermarkets is likely to be of this species. In Lake Moultrie in South Carolina individuals of up to 58 pounds have been taken. It can be caught with natural or commercial bait and with artificial lures. It has recently been introduced into China.

Channel catfish

Carolina madtom
Noturus furiosus Pl. 99
Orangefin madtom
Noturus gilberti Pl. 100
Tadpole madtom
Noturus gyrinus Pl. 101
Margined madtom
Noturus insignis Pl. 102
Speckled madtom
Noturus leptacanthus Pl. 103

The genus *Noturus* includes about 27 species known as stonecats and madtoms and occurs throughout the eastern United States and adjacent Canada. Nine species occur in the mid-Atlantic region. These small catfishes reach a maximum length of 12⅓ inches, although most species barely reach four inches. All are naked and all have four pairs of barbels, a long and low adipose fin contiguous to the upper lobe of the caudal fin (in catfishes of the mid-Atlantic region the adipose fin is higher and not contiguous to the caudal fin), and pectoral and dorsal fin spines surrounded by a sheath that contains a mild venom. The rear edge of the pectoral fin spine is usually serrate (toothed), and the size and morphology of the spines are important in identification of the various species. The spines are erectile and can be locked into position to provide an effective deterrent against would-be predators and human molesters. An attempt to pick up one of these smooth, small, wriggling fishes often results in a freely bleeding puncture wound that burns like a bee sting. Despite this drawback, because of their small size, often colorful appearance, and usually quiet nature, they make attractive specimens for the home aquarium. One species in the region, the yellowfin madtom, *N. flavipinnis,* is listed as threatened by the federal government, and one of the nine species has not yet been formally described.

Description. Carolina madtom: 1.7 to 5.2 inches (43 to 132 mm). This is a tan-colored fish with a wide black stripe on the side from the snout to the base of the tail. The two stripes are connected by four saddles across the back, and one saddle extends almost to the upper edge of the adipose fin. There are two pronounced dark crescent-shaped bands in the caudal fin. The **mountain madtom**, *N. eleutherus* (Pl. 98), is similar, but it lacks the dark band in the middle of the caudal fin, while it has a dark brown bar at the base of the caudal fin. The **yellowfin madtom**, *N. flavipinnis,* has a pale distal edge on the caudal fin and it lacks a wide black stripe along the side.

Orangefin madtom: 2.8 to 3.8 inches (72 to 96 mm). This madtom is uniformly gray-brown in color, has a pale white to orange triangle on the upper corner of the caudal fin, and has a short anal fin. The **stonecat**, *N. flavus,* is similar, but it has a cream-white spot on the dorsum just behind the dorsal fin.

Tadpole madtom: 1.1 to 5.1 inches (29 to 130 mm). The body is chubby and unicolored tan, the jaws are of equal length, and there is a faint dark line on the side.

Margined madtom: 2.2 to 5.9 inches (56 to 150 mm). The body is unicolored yellow to slate gray, there is a more or less well-defined black edge on the median fins, and the upper jaw projects slightly beyond the lower.

Speckled madtom: 1.5 to 3.7 inches (37 to 94 mm). The body and fins are tan-colored and flecked with black throughout, and the caudal fin is squared. The undescribed **broadtail madtom** (Pl. 104) is similar; but it has a more chubby body and a rounded caudal fin, and it lacks dark body specks.

Distribution and Abundance. The Carolina madtom occurs on the coastal plain and lower piedmont in the Neuse and Tar river drainages of eastern North Carolina. Because of its limited distribution, it is listed as of special concern in North Carolina. The orangefin madtom is restricted to the upper Roanoke River drainage in Virginia and North Carolina; it is considered threatened in the former state and endangered in the latter. The tadpole madtom occurs in the eastern half of the Carolinas, Virginia, and Maryland, and in all of Delaware. The margined madtom occurs in all of the mid-Atlantic region except for the southernmost portion of South Carolina, the western- and easternmost portions of North Carolina and Virginia, and the southeastern portions of Maryland and Delaware. The speckled madtom occurs in the southern half of South Carolina. Madtoms are locally common, which means that several specimens of a species may be collected at a site with high-quality habitat.

Habitat. One or more species of madtom can be found in almost any body of water in the region, be it creek, small river, or lake, with a slow or fast current, with a sand, gravel, rock, or mud bottom, and with plants or without.

Natural History. Madtoms feed on small crustaceans, insects, small fishes, and debris. They spawn from May to August. The eggs are deposited in a cluster in a shelter such as under stones or boards, or in debris such as tin cans, and are guarded by one of the parents. Some species may lay two egg clusters in a season. Records indicate that a female margined madtom contained 107 eggs and that eggs numbered 15 to 30 per cluster in the speckled madtom. The tadpole madtom reaches a length of at least 2¼ inches by the end of its first summer of life. Most species are probably mature at one year, and they rarely live longer than four years. Much remains to be learned about the natural history of madtoms in the mid-Atlantic region.

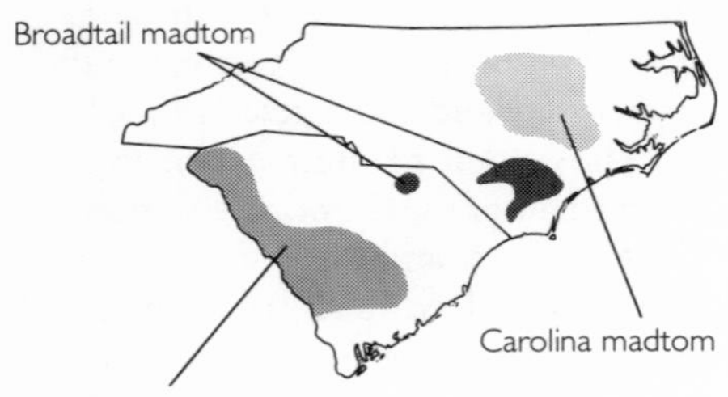

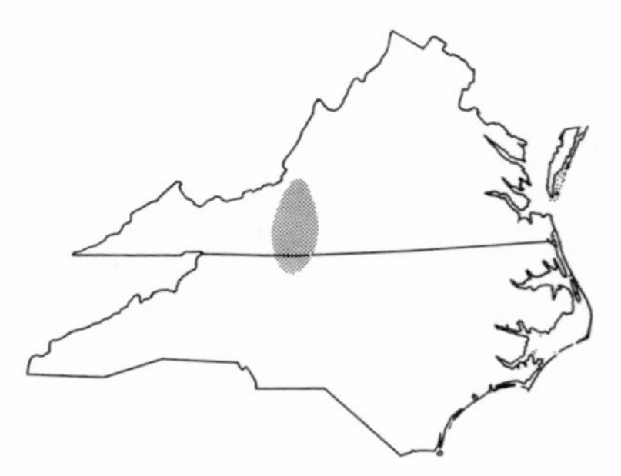
Orangefin madtom

Tadpole madtom

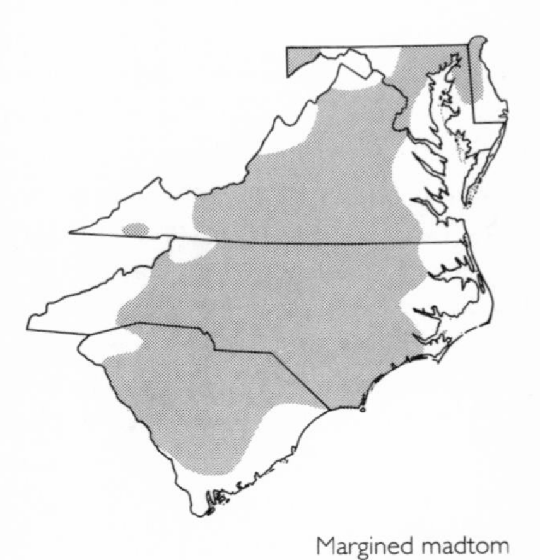
Margined madtom

Flathead catfish

Pylodictus olivaris Pl. 105

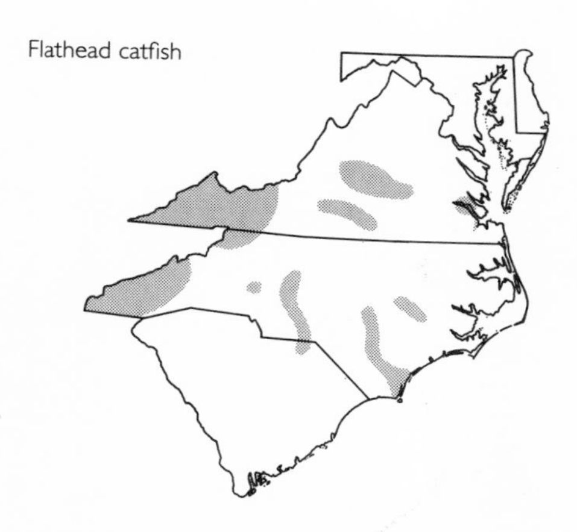

Description. 15.0 to 61.0 inches (380 to 1,550 mm). This species is separated from the other large catfishes in the mid-Atlantic region by its broad, flat head and its square tail fin, the upper tip of which is usually orange in the juvenile and white in the adult, although the white is absent in the large adult. The back is mottled yellowish-brown and grades to gray on the belly. There are fewer than 17 rays in the anal fin. The young can be separated from the similar-looking bullheads by the lateral patches of rearward-projecting teeth on its upper jaw. This species may reach a weight of more than 95 pounds.

Distribution and Abundance. The flathead catfish is native to the Mississippi River basin, and it occurs naturally in the mid-Atlantic region only in a few westward-flowing mountain streams such as the New and French Broad rivers in western North Carolina. It has, however, been widely stocked and now occurs in South Carolina in the Pee Dee, Wateree, Congaree, and Savannah rivers and in the Santee-Cooper Reservoir and in North Carolina in the Pee Dee and Cape Fear rivers. Recently it was accidentally introduced into the Neuse River, North Carolina. It is usually common.

Habitat. This catfish inhabits deep holes in large, slow rivers. It appears to prefer pools with shelter such as sunken logs or other debris.

Natural History. This species spawns in summer. The eggs are laid in a large nest fanned out on the bottom and usually adjacent to a log or other cover. The egg mass produced by one female is large and can contain up to 100,000 eggs. The male guards the eggs, as well as the fry for up to several days after hatching. Small individuals feed primarily on insect larvae and crayfishes, while larger ones eat primarily fishes. The species reaches maturity at an age of four to five years, and the maximum age that can be attained is approximately 26 years. The flathead catfish is sought by recreational and commercial fishermen. The latter take it in nets, on trot lines, and, in certain areas of North Carolina, by electricity. It is sometimes sought with heavy tackle, when it is fished for on the bottom with live or fresh bait. The flesh is considered good by many.

Pikes *Family Esocidae*

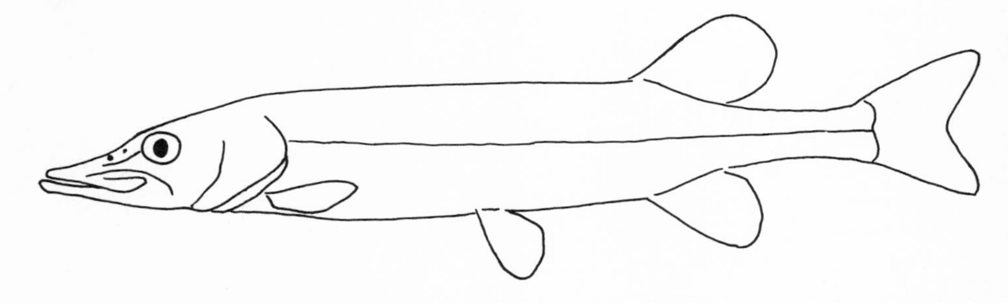

The family Esocidae includes the pickerels and pikes. The family contains one genus and five species. All species are readily identified by an elongate body, ducklike snout with sharp teeth, forked caudal fin, complete lateral line, and dorsal and anal fins that are located on the rear of the body. They lack spines in the fins. All are freshwater fishes native to northern North America, Europe, and Asia. They are voracious lie-in-wait predators that usually hide in aquatic vegetation and dart out and capture nearby prey. Three species are native to the mid-Atlantic region, and a fourth has been introduced.

Redfin pickerel

Esox americanus Pl. 106

Description. 9.8 to 15.0 inches (250 to 380 mm). The redfin pickerel rarely attains a length of more than 12 inches, and an individual that weighs more than one pound is large. It is also identified by its red pectoral and pelvic fins, vertical dark vermiculations on the side, and relatively short and wide snout.

Distribution and Abundance. The redfin pickerel is common in the coastal plain in the entire region and can be found in smaller numbers throughout much of the piedmont.

Habitat. This pickerel prefers slow streams with abundant submerged vegetation. It thrives best in the blackwater acid streams of the coastal area, but it also occurs commonly in weedy ponds and lakes. It often leaves streams to enter flooded swamps to forage in water only an inch or two deep.

Natural History. The redfin pickerel, known in much of the region as pike or bulldog pickerel, is a small but voracious predator that feeds on small fishes, frogs, and invertebrates. It is primarily a daytime feeder and waits for food in a weed bed or beside underwater shelter. It detects prey visually and catches it with a sudden, short rush. It is an early spring spawner and constructs no nest but rather lays its several thousand eggs among aquatic vegetation. There is no

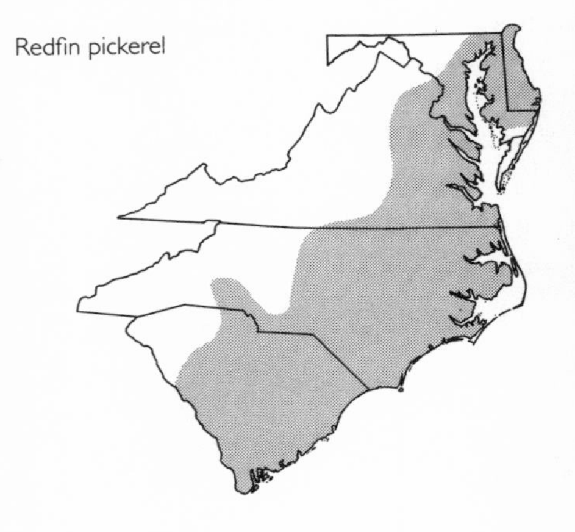

parental care of the eggs or young. The fry feeds on aquatic insects and other small invertebrates until large enough to catch fishes and larger invertebrates. Maturity is reached in one to two years, and it may live for seven or eight years.

This fish is too small to warrant the attention of most fishermen today, although a generation ago it was actively sought for food in the Carolina low country. Fishing was then done with a cane pole, a very short line, and a pork rind bait. The bait was jiggled along in shallow swamp water, and this pickerel was usually a willing taker. The redfin pickerel will also take natural bait, a bright spoon, or a spinner.

Muskellunge

Esox masquinongy Pl. 107

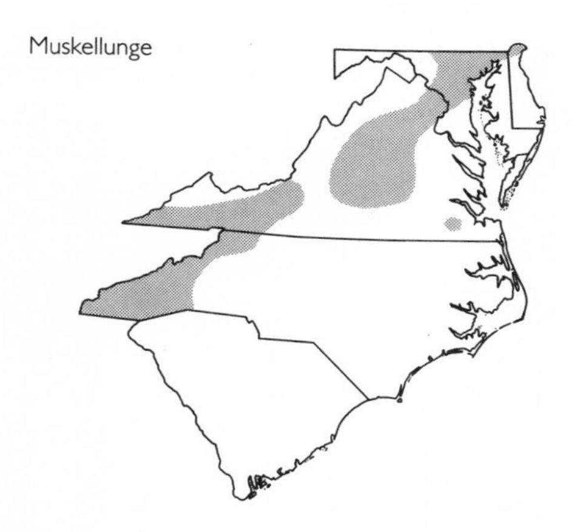

Description. 26.8 to 72.0 inches (680 to 1,830 mm). The muskellunge is the largest of the pikes and can attain a weight of nearly 70 pounds. While it usually has dark vertical bars on the body, it can be most readily identified by the scale pattern on the cheek and opercle: it has scales only on the upper half of the opercle, while the redfin and chain pickerels have scales on the entire opercle. The **northern pike**, *E. lucius*, has a fully scaled cheek and rows of spots on the body. The tiger musky is a hybrid between the muskellunge and the northern pike.

Distribution and Abundance. The muskellunge is native to the northern portion of the eastern and central United States and adjacent Canada. It occurs naturally in the mountains of the mid-Atlantic region as far south as the Little Tennessee River system in western North Carolina, although it is not known to be a native of western Virginia. It was apparently extirpated from much of its range in the region as a result of water pollution. It has, however, from 1970 into the 1980s, been reintroduced at several localities, including the Little Tennessee, Nolichucky, and New river systems in western North Carolina. It has also been widely introduced in Virginia and Maryland. It is usually uncommon to rare.

Habitat. This species prefers lakes and large rivers. It is often found in heavily vegetated areas and may occur in deep pools and open, swift water. It has been stocked most frequently in mountain reservoirs and large rivers.

Natural History. The biology of the musky is similar to that of pickerels. It spawns in the spring, builds no nest, and scatters a large number (up to 300,000) of eggs near aquatic vegetation or debris. The fry are independent at hatching. The food changes from plankton to larger invertebrates to fishes and other vertebrates as the fish grows. It reaches sexual maturity at an age of three to four years. The adult is solitary and defends a territory. Most individuals caught are from three to six years old, but the muskellunge can reach an age of at least 30 years. Spoons, lures, and live bait are used to catch it, and its large size makes it a much-sought trophy.

The northern pike has been stocked at widely scattered localities in the mid-Atlantic region. These stocking efforts have generally failed, but the fish has been successfully introduced in the piedmont and Appalachian plateau of Maryland and in a few mountain reservoirs in Virginia.

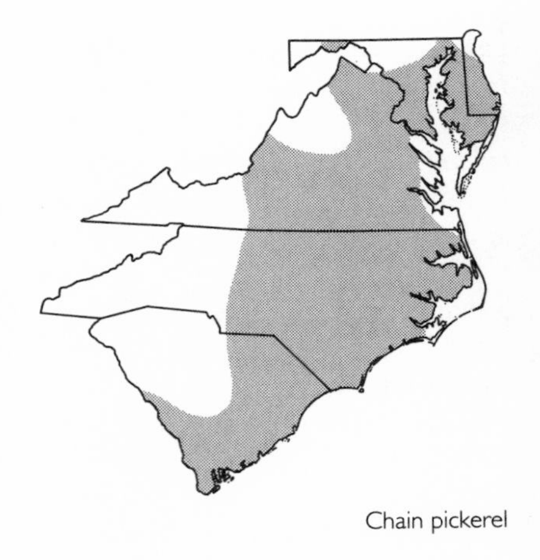

Chain pickerel

Chain pickerel

Esox niger Pl. 108

Description. 15.0 to 39.0 inches (381 to 990 mm). The chain pickerel, jackfish, or jack is identified by a mat of dark open ovals or a series of open chains on the side. At first glance easily confused with the redfin pickerel, the chain pickerel is usually larger and often reaches 18 inches in length and a weight of two to three pounds, lacks red paired fins, and has a distinctly longer and more narrow duckbill snout.

Distribution and Abundance. The chain pickerel occurs naturally throughout the coastal plain, most of the piedmont, and in some of the mountains of the mid-Atlantic region. It has been widely introduced in other portions of the region and of the United States. It is often common.

Habitat. The chain pickerel prefers the cover of aquatic plants, logs, and other debris, and it is most abundant in dark-water coastal plain streams and swamps. It is more likely to occur in lakes than the redfin pickerel, and it does not venture into the shallows as much as the redfin.

Natural History. This is primarily a predator on other fishes, although it will eat a wide variety of other animals as well. It typically lies in wait in a weed bed or next to a log, and when nearby prey is sighted, it dashes to grab it in a large mouth with many long, sharp teeth. The chain pickerel is an early spring spawner, and the female lays up to 50,000 eggs, which are scattered in aquatic vegetation and abandoned. Young fry can be found as early as March. Sexual maturity is reached by the fourth year, and it may live for eight years or more. The chain pickerel is often sought by freshwater fishermen, and it readily takes a variety of lures. Due to the presence of many small bones, it is not highly desired as food and is often released.

Mudminnows *Family Umbridae*

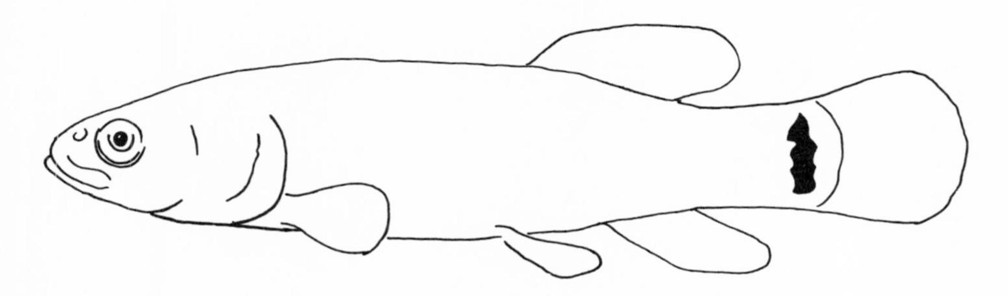

The mudminnow family contains only five species, of which four are known from the United States and Canada, the fifth from Europe. All are small, lack spines, and have a rounded caudal fin, a short snout, and a slightly flattened head. The dorsal and anal fins are set far back on the body. Mudminnows are typical of sluggish backwater habitats where there often is little dissolved oxygen. They can survive there because of a duct that connects the lunglike air bladder to the pharynx and allows them to breathe by gulping air at the water surface. Mudminnows no doubt derive their name from the behavior, of at least some species, of diving into loose mud to avoid predators.

Eastern mudminnow

Umbra pygmaea Pl. 109

Description. 2.0 to 4.3 inches (50 to 110 mm). The eastern mudminnow is easily distinguished from all other fishes in the mid-Atlantic region by its brown color and a black bar located just before the base of the rounded caudal fin. The body is robust, nearly round in cross section in the front half, and marked on the back and side with 10 to 14 thin gray to dark brown stripes with pale interspaces about as wide as the stripes.

Distribution and Abundance. This species is common to occasional throughout its range along the coastal plain from southeastern New York to northern Florida. It extends inland into the lower piedmont in Virginia and North Carolina.

Habitat. It occurs in sluggish creeks, roadside ditches, sloughs, swamps, ponds, and other habitats, where it frequents marginal areas over a mud bottom, usually with heavy vegetation and debris.

Natural History. More is known about the central mudminnow, *U. limi*, than the eastern mudminnow, and the following account is based on observations of the former species as the two are no doubt highly similar. It spawns from early spring to early summer, when, in the northern part of the range, adults may migrate from streams or ponds to backwaters of clear creeks. Sexual dimorphism is weak, but the female is slightly larger, usually lighter in color, and has an egg-distended abdomen in spring. The male swims around the female and displays his fins. He displays likewise to competing males, and torn fins suggest that males sometimes fight.

It spawns in shallow water along a heavily vegetated shore or in a nest cavity under vegetation or debris. The eggs are adhesive and are attached to a substrate or shelter in one layer. The female guards the nest and occasionally fans the eggs to ensure water circulation. Sexual maturity is reached in one to two years, and this fish rarely lives more than four. Food is aquatic insects and crustaceans. Mudminnows can easily be kept in an aquarium and fed chopped earthworms.

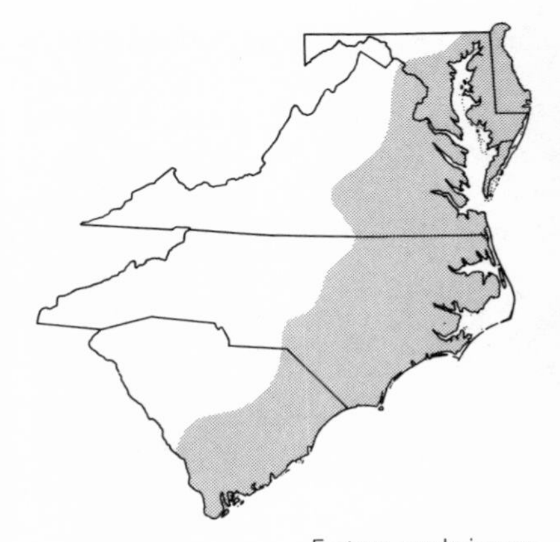
Eastern mudminnow

Trouts *Family Salmonidae*

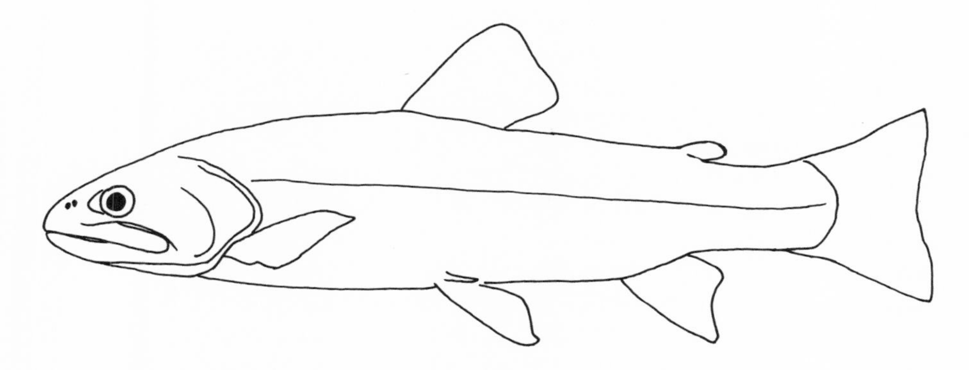

The family Salmonidae includes the salmons, trouts, chars, whitefishes, and graylings. The fishes in this family occur in fresh water or are anadromous. All are native to cold waters of the Northern Hemisphere. The majority of the almost 70 species are large, and the largest attains a length of probably over five feet and a weight of about 175 pounds; a few species are full grown at a length of only several inches. Many species are of tremendous commercial and recreational importance. Because of their size, beauty, fighting ability, and delectable taste, two of the species have been introduced worldwide. Salmonids are identified by an elongate body, an adipose fin, a single dorsal fin, paired fins placed low on the body, and a lack of spines. Salmons and trouts also have small cycloid scales loosely embedded in the skin. In addition to teeth on the jaws, they possess patches of teeth on the vomer, a bone located on the midline in the roof of the mouth. There is also an axillary process, a small triangular backward-pointing projection, located at the base of the pelvic fin. Only one species is native to the mid-Atlantic region, although two others have been introduced widely and a third locally.

Rainbow trout

Oncorhynchus mykiss Pl. 110

Description. 9.8 to 29.5 inches (250 to 750 mm). One of the two widely introduced species, the rainbow trout is named for a broad lateral stripe on the side, which is pink to red and is accentuated in the breeding adult. The dorsum is olive green and the belly whitish, and there is heavy black speckling on the entire body and all the fins. Anadromous populations of the rainbow trout, called steelhead, occur in the western United States and in other countries. The rainbow trout in the mid-Atlantic region can reach a weight of over 17 pounds. The **sockeye salmon**, or kokanee, *O. nerka,* has small black specks but no large spots on the back or caudal fin. The breeding male has a green head and a red body.

Distribution and Abundance. A native of western North America from Alaska to extreme northwestern Mexico, the rainbow trout has been widely introduced in eastern North America and on other continents. In the mid-Atlantic region it now occurs in appropriate habitat at high elevations from South Carolina northward. Populations in the region are usually ephemeral and are replenished by restocking, but the rainbow trout reproduces when habitat is appropriate. Population size varies widely with habitat, stocking, and fishing pressure. The sockeye salmon, a native of the U.S. Pacific coast, has been introduced into the Nantahala Reservoir in southwestern North Carolina, and there is apparently some natural reproduction.

Habitat. The rainbow trout occurs in creeks, rivers, lakes, and reservoirs. It is frequently stocked in ponds. It requires clean water where the temperature does not exceed 70°.

Natural History. This trout spawns in spring or early summer in the mid-Atlantic region, when it moves from lakes or pools into shallow and swift streams with a gravel or sand bottom. The female there sweeps out a depression, deposits her eggs, which are fertilized by nearby males, and then covers them with sand or gravel. There is no subsequent parental care. The fry emerge by mid-summer and feed on many small aquatic animals. The diet later shifts to larger aquatic insects and crustaceans and, for the adult, also to fishes. It reaches a length of 9 to 12 inches during its second or third year. The rainbow trout is classically fished for with a fly rod and dry or wet flies that imitate natural food. It may also be caught with spoons, spinners, natural bait, cheese, and corn. This is generally the most abundant trout in the region, primarily as a result of stocking. It is heavily fished for and is an excellent eating fish.

Rainbow trout

Brown trout

Salmo trutta Pl. 111

Description. 8.1 to 32.5 inches (206 to 826 mm). This trout can be separated from others in the mid-Atlantic region by the presence of fin spots only on the dorsal fin. The back is brownish, the side is sometimes golden, and both are slightly to heavily speckled with black and red.

Distribution and Abundance. The brown trout is a native of Europe, northern Africa, and western Asia. It has been widely introduced, and in the mid-Atlantic region occurs from northern Delaware and Maryland southward in the mountains to northwestern South Carolina. Abundance varies with stocking, habitat, spawning success, and fishing pressure.

Habitat. The brown trout occurs in small creeks, rivers, lakes, and reservoirs, under a wide range of conditions. It tolerates warmer water than other trouts but apparently does best where summer water temperatures rarely exceed 70°. It also tolerates more silt than other trouts and can survive where brook and rainbow trouts cannot.

Natural History. This trout often feeds at night, typically on insects, other arthropods, salamanders, frogs, and fishes.

Brown trout

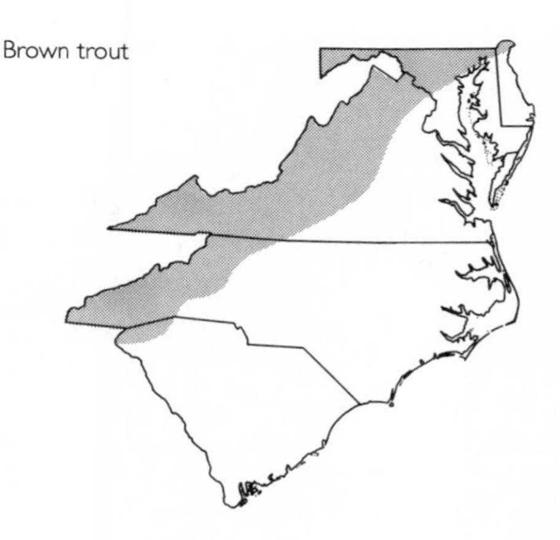

It courts and breeds in autumn, often after an upstream migration. The female makes a depression (redd) on a stream bottom, the eggs are laid there and fertilized, and the embryos develop over the winter under a layer of gravel. The fry hatch in spring and grow rapidly. Sexual maturity is reached in two or three years, and the brown trout seldom lives more than five years. Habitat for it is limited in the region.

While the brown trout is not as popular as are the more abundant, widespread, and easily caught species, this difficult-to-catch species is highly sought by a small but dedicated group of anglers, and a large one is considered a real trophy. Older brown trout feed primarily at night and fishing for them then, if legal, may be productive. Individuals can weigh more than 15 pounds.

Brook trout

Salvelinus fontinalis Pl. 112

Description. 7.9 to 20.0 inches (200 to 508 mm). This colorful trout is identified by a white margin on all but the dorsal fin. The back is dark olive with wavy markings, the side is lighter with red spots (each surrounded by a blue halo), and the lower side is orange, especially during the breeding season. The caudal fin is not forked as it is in many trouts. This small trout seldom exceeds a length of ten inches in native waters. Native trout in waters south of the New River in Virginia are a genetically different strain from those to the north. Fishermen refer to them as "speckled trout" or "speckles."

Distribution and Abundance. The brook trout is native to eastern Canada and the northeastern United States, and the southern terminus of its range is the higher elevations of the southern Appalachian Mountains. In the mid-Atlantic region it occurs as scattered populations in most of Maryland, in western Virginia, and in the mountains as far south as northwestern South Carolina. It is listed as of special concern in South Carolina. The northern strain is often stocked, and its presence and abundance vary widely and depend on environmental conditions, presence or absence of competitors, stocking, and fishing pressure.

Habitat. This trout is adapted to cold, clear, fast-flowing streams, which in the mid-Atlantic region are restricted to

Brook trout

higher elevations. It often inhabits creeks so small that a person can span one with a single step and favors creeks that are heavily shaded by dense thickets of rhododendron. It requires a high level of dissolved oxygen and cannot tolerate waters with a temperature regularly above 66° or that are fouled by silt or other pollutants. This trout also occurs in high-altitude beaver ponds. It is often stocked in ponds and occasionally in high mountain lakes and reservoirs.

Natural History. The brook trout feeds primarily on aquatic insects and other arthropods. Hatchery-raised individuals will readily take canned corn. It is usually a diurnal feeder, but when the temperature is high, it feeds at night. Stream-dwelling brook trout breed in early to mid-fall in the mid-Atlantic region, when they migrate upstream to find a section of clear and cold creek with a clean gravel bed. Here the female digs a redd or nest several inches in diameter and an inch or two deep, which the male defends against intruders. When the spawning area is ready, the female deposits from 100 to 5,000 eggs, which are then fertilized. The sticky eggs adhere to gravel in the redd and are covered with more gravel by the female, after which the adults abandon the site. The developing embryos are bathed in clean, cold, oxygen-rich water and hatch early the next spring, when the fry work their way to the surface. Maturity is reached at an age of two to three years.

Growth is slow in its southern mountain stream habitat, and the adult is often only six to ten inches long. It is, however, eagerly sought by the avid fisherman who appreciates the solitude and beauty of these small isolated brooks and the feisty nature of this colorful native. The brook trout grows much larger in trout ponds, but fishing for it there lacks the appeal of seeking a true wild trout in one of the last remnants of the eastern mountain wilderness.

Pirate Perches *Family Aphredoderidae*

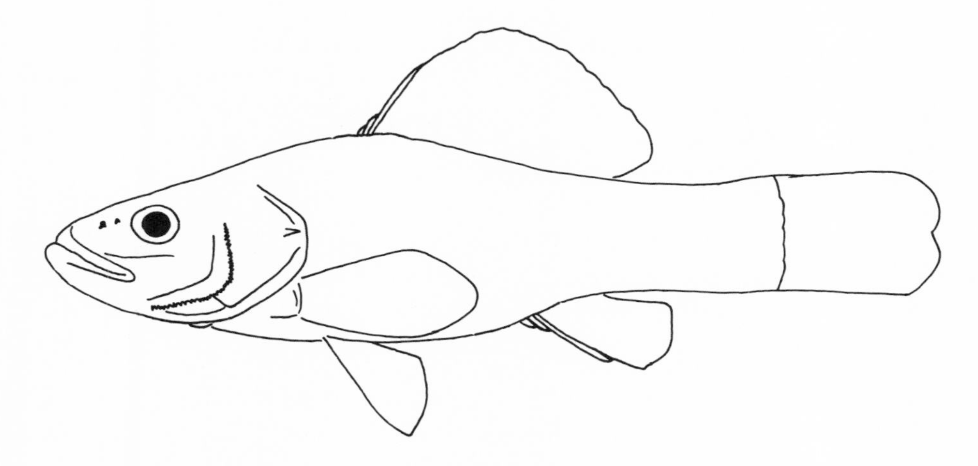

Strictly a freshwater species, the pirate perch is the only living member of its family. The description and characteristics of the species thus defines the family.

Pirate perch

Aphredoderus sayanus Pl. 113

Description. 2.5 to 5.7 inches (64 to 144 mm). This is a stout-bodied, small, chocolate brown fish with a single dorsal fin and a large head and mouth; there is often a purplish sheen on the lower side. There is a dark bar below the eye and a larger one just before the base of the tail. The dorsal and anal fins have two or three weak spines. The anus in the adult is located in the throat area.

Distribution and Abundance. The pirate perch is found in the coastal plain and the piedmont of the Atlantic and Gulf slopes and in much of the Mississippi River basin. Although frequently encountered near the coast, it is rarely taken in numbers.

Habitat. It is found in swamps, backwaters, creeks, roadside ditches, and other areas of slow water, soft bottom, and abundant aquatic vegetation.

Natural History. The pirate perch was named by C. C. Abbott, an ichthyologist of the mid-1800s, who observed it to eat only fishes in an aquarium. In nature it eats predominantly larvae of aquatic insects and small crustaceans. It hides in dense vegetation or bottom debris during the day and feeds at night, primarily just after dark and before dawn. It breeds in early spring, and it

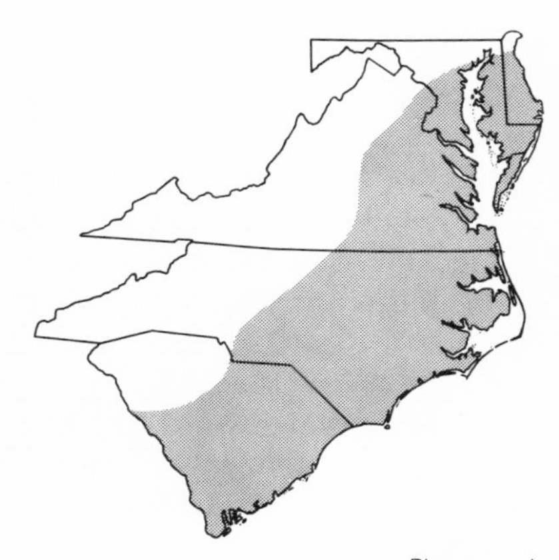

Pirate perch

has been noted to build and guard a shallow nest depression about two inches wide. The female produces between 129 and 160 eggs per spawning, which are deposited mostly on the periphery of, or in, the pit. The young grow rapidly after hatching and can attain over two inches in their first year. Few live longer than four years. In the very young the anus is located in the "normal" position, meaning just anterior to the anal fin, but it "migrates" forward as the fish grows. *Aphredoderus* means "excrement throat," and refers to the position of the anus in the mature fish.

Cavefishes *Family Amblyopsidae*

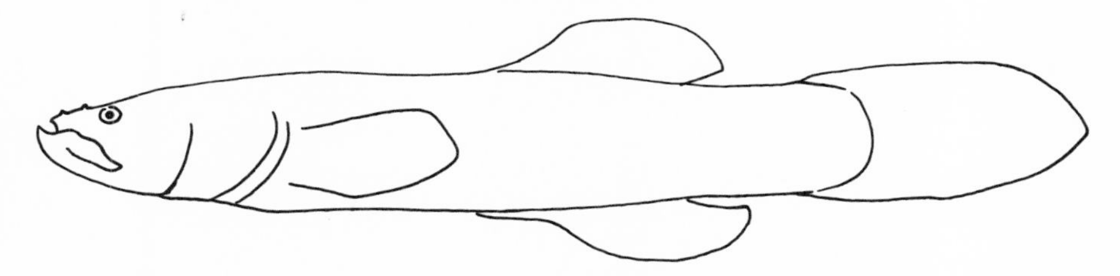

This family is restricted to fresh water in the regions of the eastern United States not previously glaciated. There are six species, four of which are blind and only two of which have functional eyes. Only one species is found in the mid-Atlantic region. The group has cycloid scales and a naked head, and all except one species lack pelvic fins. The largest species can reach a length of about 3½ inches.

Swampfish

Chologaster cornuta Pl. 114

Description. 0.9 to 2.7 inches (23 to 68 mm). The swampfish is small, has tiny eyes, and lacks pelvic fins. It is bicolored, and the brown back contrasts sharply with the creamy white belly; there are three dark stripes on each side. The mouth is large, and the male develops a fleshy protuberance on the upper jaw. As the fish grows, the anus migrates forward to its location on the underside of the head.

Distribution and Abundance. The swampfish occurs in the coastal plain from southeastern Virginia to Georgia. Although locally common, it is taken only uncommonly or rarely, probably because it is largely nocturnal and because of the heavily vegetated habitat in which it lives.

Habitat. This species prefers acidic, tannin-stained, calm waters such as swamps, roadside drainage ditches, ponds, and sluggish creeks, where it is usually associated with vegetation, particularly *Sphagnum* moss. All other members of this small family are adapted to a subterranean life and are found in caves in limestone areas just to the west of the region.

Natural History. The swampfish eats small crustaceans such as scuds and water fleas and aquatic insects such as midges. It feeds mostly at night. It spawns in March and April, but details are not known. One cave-dwelling member of this family incubates the eggs in the gill cavity. The swampfish probably does not do this, however, because, although the female carries few eggs, there are too many to fit into the gill area and no swampfish has ever been noted to so carry its eggs. The young grow rapidly after they hatch, and some individuals may live up to two years, although only a few exceed an age of 15 months.

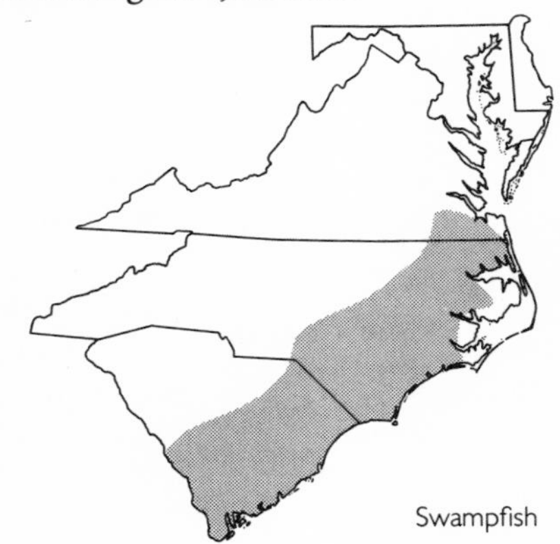

Swampfish

Topminnows and Killifishes *Family Fundulidae*

This group has recently been separated from a much larger group (the Cyprinodontidae) of small freshwater, brackish water, and coastal marine fishes found in a wide range of habitats through much of the temperate and tropical parts of the world. The fundulids include about 40 species found from southeastern Canada through the eastern half of the United States south to the Yucatan Peninsula in Mexico, as well as in Bermuda and Cuba. The Fundulidae contain some of the smallest fishes in the world. Species in the mid-Atlantic region have an elongate body that is oval in cross section, paired fins that are placed low on the body, and dorsal and anal fins that are located far back on the body; all fins lack spines. All species have a small terminal or superior mouth.

Many species occur in extremely variable environments, such as salt marshes, and their adaptations to such environments have resulted in some being widely used as bait fishes owing to their extreme tenacity to life. Because of the abundance of many species, they are valuable forage fishes. They are especially valuable to man in the control of insect pests, especially mosquitoes. Because of their small size, attractive colors, interesting behaviors, rapid growth and maturation, and the ease with which they can be maintained in captivity, many species have become popular aquarium fishes and many are used as subjects for the study of problems in basic fish biology and ecology.

Northern studfish

Fundulus catenatus Pl. 115

Description. 3.0 to 7.1 inches (76 to 180 mm). This fish has a yellow-brown back and a silver-blue side. The breeding male is quite attractive, with a bright blue side, red spots or lines on the head and fins, yellow fins, and a black bar on the tail. There are eight to ten rows of small brown spots on the side of the young and on the adult female, while on the male these dots are red-brown. The male is larger than the female.

Distribution and Abundance. In the mid-Atlantic region it is found only in the Tennessee River drainage of far western Virginia, where it is the only killifish that occurs. It is uncommon to occasional.

Habitat. The northern studfish inhabits calm areas of creeks and small rivers, typically in shallow sandy backwaters of clear and rocky creeks. It often lies in or near vegetation.

Natural History. This species is often seen in small groups that prowl the

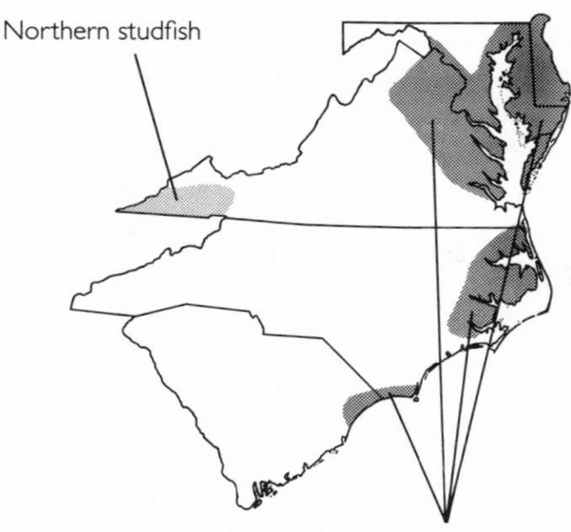

shoreline for food or that rest in backwaters. It is largely protected from fish predators by the shallow water in which it lives and from bird and mammal predators by its speed and by finding cover in aquatic plants. Food is often obtained on the bottom and consists of the young and adults of many insects, beetles, snails, clams, worms, and crayfishes. It spawns from mid-May to early August. The male guards a territory near the shore, and the eggs are deposited on clean gravel.

Golden topminnow

Fundulus chrysotus Pl. 116

Description. 1.4 to 3.3 inches (35 to 84 mm). This is an attractive pale olive green or yellowish fish with conspicuous small iridescent gold or greenish, and pearl white, flecks on the side. The adult male often has 8 to 11 faint greenish bars on the side, which are absent in the female. The breeding male develops reddish flecks on the rear half of the body and on the median fins, and the fins become yellowish with some reddish-orange color. The reproductive female becomes yellowish and develops a black eye but not the red body flecks and yellowish fins.

Distribution and Abundance. In the mid-Atlantic region this species is found only in the coastal counties of South Carolina, from Georgetown southward. It is widely distributed in the southeastern coastal United States from Georgia to eastern Texas and in the lower Mississippi River basin. It is usually common in southeastern South Carolina.

Habitat. The golden topminnow occurs in quiet, slow, shallow, heavily vegetated waters of marshes, swamps, lake shores, sloughs, drainage ditches, and creek backwaters. It is occasional in brackish waters.

Natural History. Food consists of a wide range of small aquatic invertebrates such as insects, mites, small crustaceans, and small snails. The golden topminnow also eats terrestrial insects that blow into or fly low over water, and small plants such as watermeal, *Wolffia* spp. It reproduces from spring through summer. The eggs are deposited singly in submerged vegetation and hatch in about two weeks. The characteristic gold spots first appear at an age of about 30 days, and when three months old the sexes can be distinguished by color. It first becomes reproductive in the summer after it hatches. Some may live long enough to see a third summer. This attractive small fish is well suited to the aquarium.

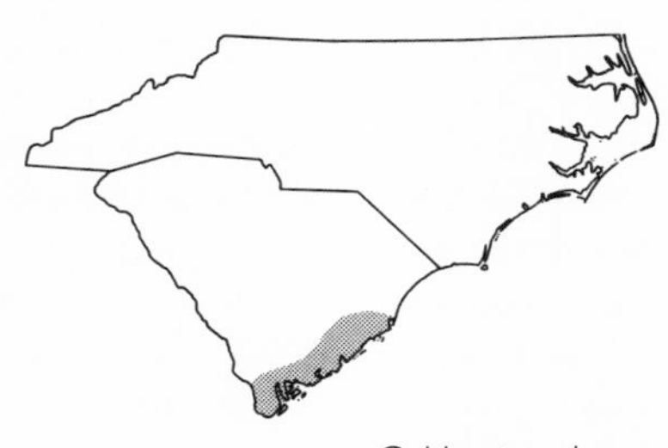

Banded killifish

Fundulus diaphanus Pl. 117

Description. 2.0 to 5.1 inches (50 to 130 mm). This species is distinguished from its relatives by a more elongate body, longer and more flattened snout, small scales, and numerous narrow bars on the side. These bars are darker than the ground color in the female and silvery in the male. The female is olive above, silvery white on the lower side, and white on the abdomen; the dorsal, caudal, and pectoral fins are yellow, the anal and ventral fins are colorless, and there are 16 to 20 greenish bars on the side. The male is colored as the female, but the back is more greenish, the dorsal and caudal fins dusky, the anal and pectoral fins yellowish, and the ventral fin bluish white and tinged with yellow. The side is often iridescent. The young is colored as the female. The adult female is longer than the male.

Distribution and Abundance. Within the mid-Atlantic region, this killifish is found in Maryland and Delaware, the northern half of Virginia and all of the coastal area of that state, and south along the coast to near Georgetown, South Carolina (see range map for northern studfish). It also occurs widely in the northeastern United States and west to eastern Montana. It is often common to abundant.

Habitat. The banded killifish prefers calm, slow, clear or brown-stained waters of lakes, ponds, rivers, and creeks with a sand and gravel bottom near scattered submerged and emergent vegetation. It is often found in tidal fresh water, and it is moderately rare in brackish water (salinity of up to 20 parts per thousand).

Natural History. This killifish occurs in large and loose schools. Food is any small invertebrate it can obtain: crustaceans, insects, mollusks, and annelid worms. Unidentified vegetable matter has also been found in its stomach. It feeds at any level in the shallow water column. It spawns only during the day. A male establishes a loosely defined territory of about 16 square feet and vigorously courts ripe females; he has been observed to continue courtship after entering the territory of another male. Fertilization of eggs has been observed in water six to eight inches deep, typically over filamentous algae or submerged vascular vegetation, and the eggs are attached to plants singly or in twos or threes. A large ripe female can contain 200 to 250 eggs, and ripe females have been collected from April through September. The eggs hatch in about nine days, the exact time depending on the water temperature, when the young are less than one-half inch long. Black bars appear about ten days after hatching, and the sexes can be distinguished externally when about three months old. The banded killifish probably can reach an age of at least three years. It is an excellent bait fish and also a fine aquarium fish if a larger tank is available. No doubt this species is an important forage fish as well as a controller of noxious insects, especially mosquito larvae.

Lined topminnow

Fundulus lineolatus Pl. 118

Description. 1.3 to 3.4 inches (33 to 86 mm). This elegant fish has a large gold spot on the top of the head, a conspicuous black bar under the eye, and a dirty white side on which there are six to eight blackish body-length stripes in the female and 11 to 15 blackish bars in the male. The male and female are of about the same length.

Distribution and Abundance. In the mid-Atlantic region it occurs from extreme southeastern Virginia south along the coastal plain of the Carolinas. It also occurs further south. It is often common.

Habitat. The lined topminnow is an active and alert fish at the surface of quiet waters such as sloughs, drainage ditches, ponds, lake edges, and stream backwaters, especially near submerged or emergent vegetation. It prefers waters that are soft, low in nutrients, acidic, and clear, although these are often tea-colored from dissolved tannins.

Natural History. The lined topminnow spawns in mid-spring and probably through the summer when the male establishes a territory that may measure six to eight feet along a shore in water one to two feet deep. There is a distinct courtship; the eggs are probably laid singly, each is about one-sixteenth inch in diameter, and they hatch in 10 to 14 days. The young are slender and transparent, and they swim at the water surface. Food consists of a range of small larval and adult invertebrates such as insects, snails, crustaceans, and mites, as well as minute floating plants. It makes an attractive aquarium fish. In some areas it is sought as a bait fish.

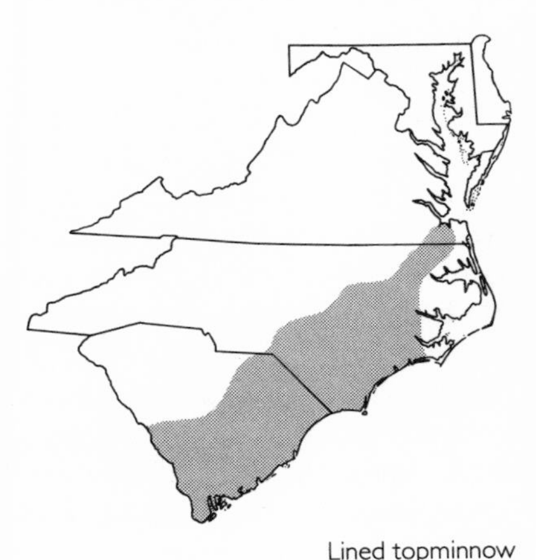

Lined topminnow

Speckled killifish

Fundulus rathbuni Pl. 119

Description. 1.9 to 3.8 inches (48 to 96 mm). This fish is characterized by a black line on the side of the head from the mouth to the eye, a yellow-brown back and side, and a white to yellow underside. The black line on the side is absent in the juvenile. There are numerous dark brown spots on the back and side of the juvenile and female, but these are absent in the male. The fins are clear to yellow. The breeding male is golden brown with black spots on the side of an iridescent gold head. He can develop an iridescent pale blue sheen on the sides of the body.

Distribution and Abundance. The speckled killifish is restricted to the mid-Atlantic region, where it is found in central North Carolina and adjacent south central Virginia. It is locally common.

Habitat. It occurs in backwaters, pools, and inlets of creeks and smaller rivers, usually over a bottom of mud or sand.

Natural History. Little is known about this fish. It spawns from at least mid-May through July. The female contains some 300 eggs of different size classes, which indicates that not all are deposited in one season or that they are deposited over a prolonged period in one season. In one observation, young measured on 1 August were slightly over a half inch long and by the end of Sep-

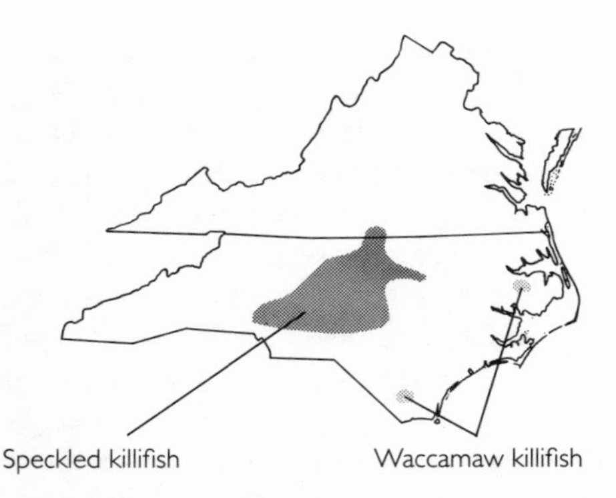

tember were about one inch long. Most young are probably not sexually mature until the second summer of life. The speckled killifish feeds on the bottom, on a wide range of larval and adult insects; a young shiner was found in one stomach.

Waccamaw killifish

Fundulus waccamensis Pl. 120

Description. 2.0 to 3.9 in (52 to 100 mm). This fish is similar to the banded killifish, but it is more slender.

Distribution and Abundance. It is restricted to North Carolina, where it is known only from Lake Waccamaw in Columbus County and Lake Phelps in Washington and Tyrrell counties (see range map for speckled killifish). Because of its limited distribution, it is listed by the state as of special concern.

Habitat. This midwater species occurs in large schools over a sandy bottom away from shore and near emergent vegetation on shoals or near shore. It is common in winter in swamps and canals around Lake Waccamaw.

Natural History. The Waccamaw killifish feeds on small bottom-dwelling insects and crustaceans. The male defends a territory and spawns over a sandy substrate with passing females. Approximately 100 eggs are deposited per spawning. The spawning season extends from early March to August.

Bluefin killifish

Lucania goodei Pl. 121

Description. 0.7 to 1.9 inches (19 to 49 mm). This species looks like a *Fundulus* that has been laterally compressed: it has a relatively deep and thin body. There is a black stripe on the side from the snout through the eye to the base of the tail; the body is grayish white. The dorsal and anal fins of the male contain much blue and are edged with black; the fins of the female are clear.

Distribution and Abundance. In the mid-Atlantic area it is found only around Charleston, South Carolina, and in Wilmington, North Carolina, both of which are large population centers for humans—a fact suggesting that it may have been introduced to both areas. The bluefin killifish is listed as of special concern in North Carolina. It is also found in coastal Georgia, in all of Florida except the western panhandle, and in extreme southeastern Alabama. It is locally common.

Habitat. The bluefin killifish occurs in calm waters of drainage ditches, sloughs, ponds, and backwaters of creeks and rivers, in or near dense aquatic vegetation. It is typically found in fresh water, but it also occurs in moderate salinities.

Natural History. In South Carolina it probably breeds from spring to summer, while further south it probably breeds throughout the year. The male establishes a territory that is centered around bottom-growing vegetation. He constantly patrols his domain, chasing off rival males and courting females that wander in. The bright median fins are then quickly raised and lowered in a behavior called "fin-flicking"; the rapidly concealed and revealed bright colors make the fish more conspicuous than if the bright fins were kept constantly raised. It spawns against submerged vegetation, and the eggs are probably deposited singly. It does not live longer than two years. The food is primarily tiny plants and animals that grow on larger plants, as well as bits of vascular plants such as tapegrass, *Vallisneria* sp. It makes an interesting aquarium fish.

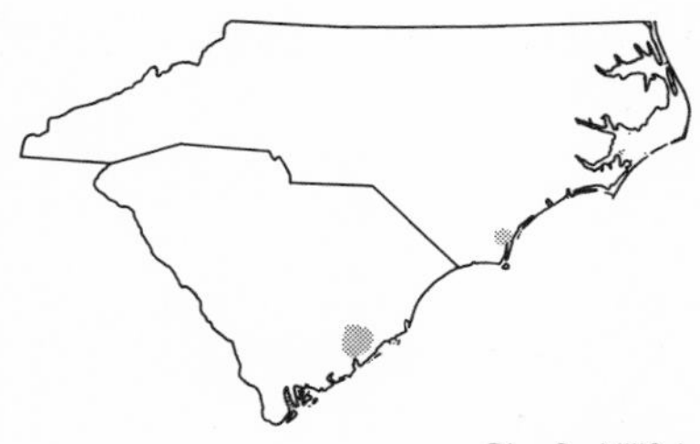
Bluefin killifish

Livebearers *Family Poeciliidae*

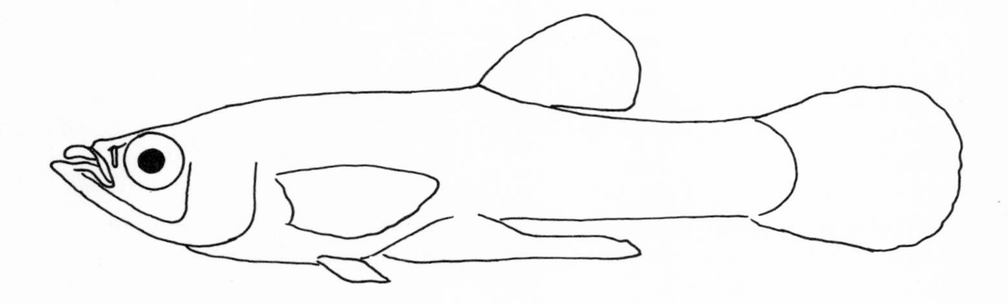

Members of this family give birth to live young, except in one species, and the group occurs from southern New Jersey south throughout the eastern United States, on both coasts of Mexico, through Central America, and in South America down to northeastern Argentina; it also occurs on most islands of the Caribbean. This family contains about 150 species and includes the world-famous guppy, swordtails, platys, mollies, and other extremely popular aquarium fishes. Sperm is transferred directly into the female via the modified elongate rays of the anal fin (gonopodium) of the male. The group occurs in fresh and brackish waters at low elevations, usually in shallow, quiet, and heavily vegetated areas. Poeciliids usually feed on small invertebrates at or near the water surface, although some are primarily, or exclusively, herbivores. This group contains some of the world's smallest fishes, and the word "guppy" has become a synonym for small. Three poeciliids are found in the mid-Atlantic region.

Eastern mosquitofish

Gambusia holbrooki Pl. 122

Description. 0.8 to 2.6 inches (20 to 65 mm). This is a small gray fish with a black bar below the eye, a head flattened on top, a large eye, one to three rows of black spots on the dorsal and tail fins, and scales edged with black. The latter produces a cross-hatched appearance, especially on the upper side. Spines are absent. The central rays of the anal fin are greatly elongated in the male and form the gonopodium. The female typically has a prominent large dark spot on the body near the urogenital opening. Occasional males are melanistic (black). The male reaches a length of 1 inch, the female 2½ inches. Populations in the mid-Atlantic region were long known commonly as mosquitofish and scientifically as *G. affinis*, but it has recently been determined that there is an eastern mosquitofish, *G. holbrooki*, and a similar-appearing western mosquitofish, *G. affinis*. There is a possibility that both species occur in the region.

Distribution and Abundance. In the mid-Atlantic region this species is found throughout much of both Carolinas, although populations are scattered and local in the foothills. Further north it becomes progressively more restricted to the coast, and in northern Maryland and Delaware it occurs only on the edges of the Chesapeake and Delaware bays. The northeastern terminus of its natural range is probably Delaware, and the populations in New Jersey were

Eastern mosquitofish

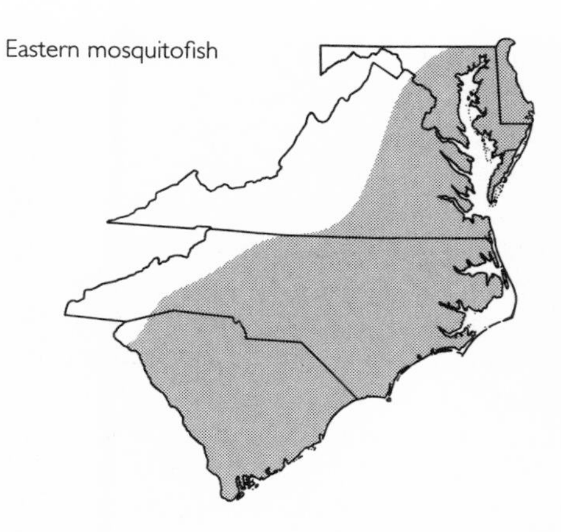

probably introduced. It also occurs south through all of Florida and west into the Mobile River drainage in Alabama. It is common to abundant in slow and calm fresh waters and also inhabits estuarine waters. As a result of introduction for mosquito-control purposes, it and the closely related western mosquitofish are now found through much of the United States and northern Mexico.

Habitat. The eastern mosquitofish favors vegetated areas of lakes, oxbows, ponds, drainage ditches, sloughs, and backwaters of creeks and rivers over a soft substrate such as mud or mud/sand. It can tolerate waters as warm as 104°. It is often found in degraded waters from which other natives have been eliminated.

Natural History. This is often the most conspicuous fish in its habitat as it swims energetically to and fro at or just under the water surface while it searches for food and, in the case of the adult male, for a female. The male courts almost continuously and fertilizes females throughout the warm months. The gestation period is 21 to 28 days or less, and three or four broods may be produced in one season. The female contains from 1 to 315 embryos. The young rapidly reach sexual maturity, often in their first year. Like many small and tiny fishes, few survive beyond their first year. Its food consists primarily of surface-dwelling aquatic insects and their larvae, small crustaceans, algae, and its own young. In disturbed habitat it often replaces other fishes. It makes an enjoyable, instructive, hardy, and inexpensive aquarium fish that can be kept in a community tank.

Least killifish

Heterandria formosa Pl. 123

Description. 0.6 to 1.4 inches (14 to 36 mm). This is the smallest fish in the mid-Atlantic region and one of the smallest in the world, with the male rarely longer than three-quarters of an inch and the female one inch. It is similar in appearance to many of the killifishes, hence its common name. It also looks like the mosquitofish but is smaller, more colorful, and more rounded in build. The eye is large, the top of the head is flat, and the mouth is located at the top of the short snout. The body is robust anteriorly and then abruptly reduced posteriorly at midbody. There is a large dark spot on the dorsal and anal fins of the female and on the dorsal fin of the male. A dark stripe on the side extends from behind the eye to a dark spot near the base of the tail, and six to nine dark bars of irregular length cross the stripe. The back and side are brown or gray, and the belly is white. The dorsal fin is short and low and similar in size and shape to the anal fin in the female, while in the male the first few anterior rays of the anal fin are extended and form a gonopodium.

Distribution and Abundance. In the mid-Atlantic region the least killifish is found only along the coast from Wil-

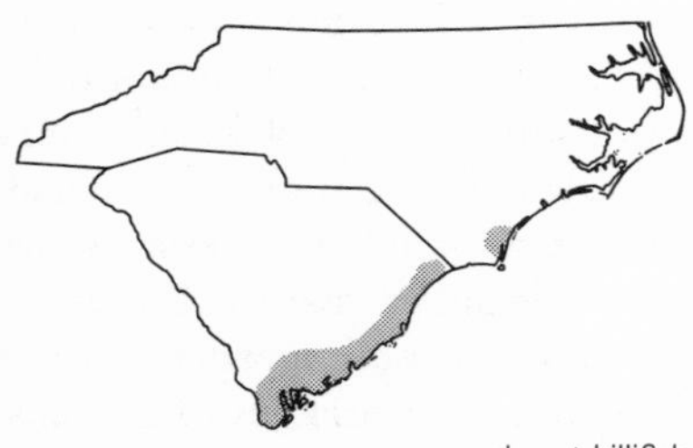
Least killifish

mington, North Carolina, to the south. It also occurs along the coast of Georgia, throughout all of Florida, and along the Gulf coast in Louisiana almost to Texas. It is usually common to locally abundant. It is considered to be a species of special concern in North Carolina because of its limited distribution and because of the few specimens known from there.

Habitat. This fish inhabits ponds, sloughs, drainage ditches, and stream backwaters, where it occurs in shallow, heavily vegetated, and soft-bottomed areas. Although found primarily in fresh water, it also occurs in waters with a salinity of up to 30 parts per thousand.

Natural History. Like the mosquitofish and many killifishes in its range, the least killifish is an omnivore and is usually observed as it busily searches the water surface for tiny aquatic and terrestrial invertebrates, detritus, algae, and vascular plants. It reproduces from early spring to late summer. The eggs are fertilized while in the body of the female, via the gonopodium of the male; mating lasts only for a second. After a gestation of three to four weeks, the female gives birth to from one to eight young at intervals of 3 to 41 (with an average of about 10) days between broods. Up to six stages of developing young may be present in a female simultaneously. Apparently as the young that result from one group of fertilized eggs are produced, new eggs mature and are fertilized, and this results in a constant "rollover" in production and fertilization of eggs and birth of young, a phenomenon known as superfetation. A long-term average of about one young per day per female is produced. Many young reach sexual maturity in their first year of life and, in fact, few survive it.

The least killifish, because it is colorful, lively, and can readily be kept with other species, is an ideal aquarium fish. Its small size makes it possible to maintain a population in a tank that could hold only a few individuals of a larger species. As with all fishes, much of interest can be learned about it by careful observation and recording of information.

Sailfin molly

Poecilia latipinna Pl. 124

Description. 0.7 to 5.9 inches (18 to 150 mm). This relative of the mosquitofish and the least killifish is identified by its larger size, large saillike dorsal fin which begins at the anterior portion of the dorsum, a deep compressed body, and a very deep caudal peduncle. About five rows of dark brown spots are present on the body. The edge of the dorsal fin in the male is orange, this fin is higher than in the female, and there are some five vertical rows of spots on the dusky-edged tail fin. Captive breeding has resulted in a proliferation of a number of morphs, such as black, yellow, pied, and silver individuals, and those with sickle-shaped median fins.

Distribution and Abundance. In the mid-Atlantic region this fish occurs along the coast from the lower Cape Fear River, North Carolina and to the south. It occurs further south on the Atlantic Coast and along the Gulf of Mexico coast to Mexico. It has been widely introduced in the western United States and Canada and in other countries. In the mid-Atlantic region it is probably uncommon.

Habitat. In the Carolinas it is known only from cordgrass marshes in brackish water. In Florida and farther west it is a coastal as well as a freshwater form, and in the latter it occurs in calm, shallow, warm, and heavily vegetated waters such as ponds, roadside sloughs, lake margins, and stream edges and backwaters.

Natural History. The sailfin molly feeds on algae and vascular plants, detritus, and mosquito larvae. In Florida, where it is abundant, it is often seen at the surface, along with the related eastern mosquitofish, as it busily searches for food and mates. Courtship and copulation are rapid, sperm can be stored by the female, and a large female can produce up to 141 young, each just under 1/2 inch long, in one brood. Growth is rapid and it probably lives for one or two years. It is little known in the mid-Atlantic region. It is a popular and easy to maintain aquarium fish.

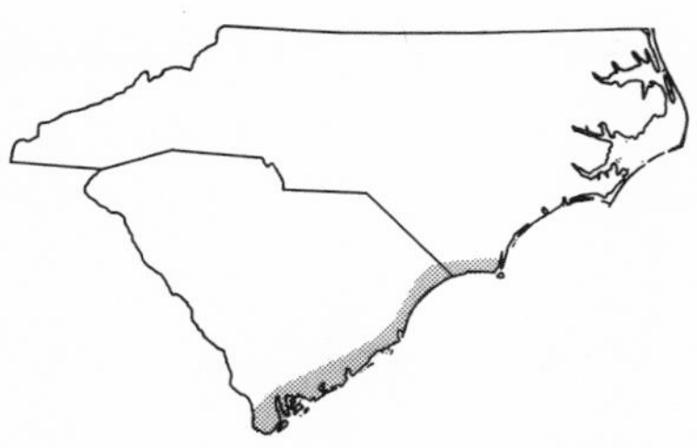

Sailfin molly

Silversides *Family Atherinidae*

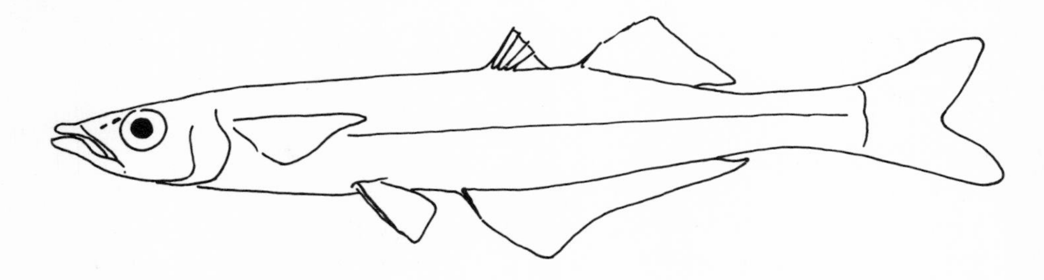

Silversides are small, thin-bodied, translucent fishes with two widely separated dorsal fins and a sickle-shaped anal fin. They superficially resemble minnows (family Cyprinidae), but the latter have only one dorsal fin. They occur worldwide, and there is a total of about 160 species. They are predominantly marine in tropical and subtropical waters. Three species occur in the fresh waters of North America, and all three are found in the mid-Atlantic region. Silversides swim in large schools near the water surface and are an important food for larger fishes. The common name refers to a distinct silver stripe on the side, which often reflects sunlight like a mirror when these fishes turn near the water surface.

Brook silverside

Labidesthes sicculus Pl. 125

Description. 2.5 to 5.1 inches (66 to 130 mm). The body is highly elongate and the snout is long and pointed (beaklike). There are from 74 to 87 lateral scales. In life, the air bladder is visible through the skin and appears as an oblong bubble near the center of the body. The breeding male develops a red snout and a black tip on the first dorsal fin.

Distribution and Abundance. This fish occurs in the mid-Atlantic region only in South Carolina, where it is abundant in most major drainages, and in extreme southwestern Virginia, where it is rare. It is also native to the Mississippi River basin and the Great Lakes basin and to the Gulf coastal plain.

Habitat. The brook silverside is a surface fish, occurring in the calm portions of creeks, lakes, and small rivers.

Natural History. This is a highly active freshwater fish. It feeds during the day on midge larvae, insects, and zooplankton, which it takes near the surface, and can also catch flying insects in the air and strain small organisms from the water with its fine gill rakers. During dark nights it lies suspended motionless in the water, but during the light of a full moon it dashes, splashes, circles, and leaps out

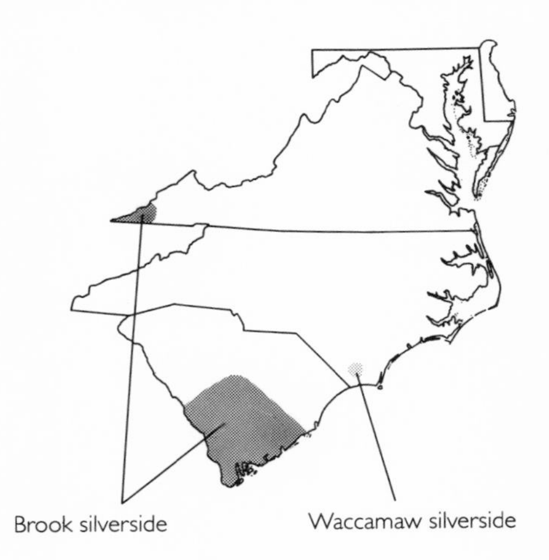

of the water—and indeed is widely known as "skipjack" because of these graceful arching leaps. Leaps play a role in feeding, predator avoidance, and spawning and may be just plain fun. The brook silverside can be attracted to a beam of light at night and will rapidly follow a moving beam.

It spawns in spring and summer. The male then actively pursues a female and fertilizes the eggs while they are still inside the body of the female. The details on how he does this are not known. The female later releases the fertilized eggs, which adhere to vegetation or other objects. After they hatch, the fry move from shallow to deep water in compact schools. Adult size is reached within three months after hatching. Most populations are annuals, meaning that an entire life cycle is completed within one year. Like other silversides, the brook silverside is an important forage fish for predatory fishes as well as for birds, turtles, snakes, and mink.

Although all silversides make an unusual addition to an aquarium, they are difficult to keep because they do not tolerate handling and because they require live food, such as larval or adult brine shrimp. Silversides can best be collected alive by corralling them in a seine net and then dipping them into a small container of water without touching them. Silversides are not good bait fishes because they survive poorly in a bait bucket.

Inland silverside

Menidia beryllina

Description. 2.2 to 3.9 inches (55 to 100 mm). The inland silverside has a heavier build and fewer scales (36 to 44) in the lateral line than the similar brook silverside. The body is pale yellow or green above a lateral silver stripe and white below.

Inland silverside

Distribution and Abundance. It occurs on the outer coastal plain of the mid-Atlantic region, where it is common and occasionally abundant.

Habitat. It prefers large freshwater or brackish rivers and creeks and occurs in clear, open water, often near the surface, over a sand or gravel bottom.

Natural History. This fish moves toward shallows at night and deeper waters in the day. Its diet is insects, copepods, and scuds (amphipods). It spawns in the morning during spring and summer, when the female produces eggs daily; a large female may produce 2,000 eggs per day over a three-month breeding period, and these are typically attached to vegetation. It is an important forage fish.

Waccamaw silverside

Menidia extensa Pl. 126

Description. 1.2 to 2.6 inches (30 to 66 mm). This is a pronouncedly slender, semitransparent fish with a silvery stripe on the side, a dusky back, and 44 to 50 lateral scales.

Distribution and Abundance. The Waccamaw silverside is endemic to Lake Waccamaw in Columbus County, North Carolina, and it was collected outside the lake only after high water swept some individuals over a concrete dam at the lower end of the lake and into the Waccamaw River (see the range map for the brook silverside). It is one of the most abundant forage species in Lake Waccamaw.

Habitat. It is found throughout Lake Waccamaw, where it is common in schools near the surface in open water.

Natural History. This fish is known as "skipjack minnow" or "glass minnow" by fishermen at Lake Waccamaw, as it, as well as other silversides, commonly "skips" over the water when chased by predators or when spawning. The Waccamaw silverside feeds on tiny plankton, most commonly water fleas, *Daphnia* spp. Although its reproductive behavior has not been documented, it probably spawns in large frenzied schools in open water or near emergent vegetation such as beds of maidencane, *Panicum hemitomon*, which is common near the lower end of the lake. The peak of spawning is during lake warming in spring. The female then produces a large number of eggs, which adhere to the sandy bottom or to vegetation. Both sexes mature after their first winter, and most individuals die after their first spawning season. Although often plentiful, it is a poor bait fish because it does not survive handling.

The Waccamaw silverside has a potential for rapid extinction, for if lake water quality were to deteriorate and it could not reproduce for even one year there would be few survivors to reproduce in the subsequent year. Because of its restricted distribution and because of the possibility of rapid lake water quality deterioration as a result of nutrient overloading, the Waccamaw silverside has been designated as threatened by the federal government.

Sticklebacks *Family Gasterosteidae*

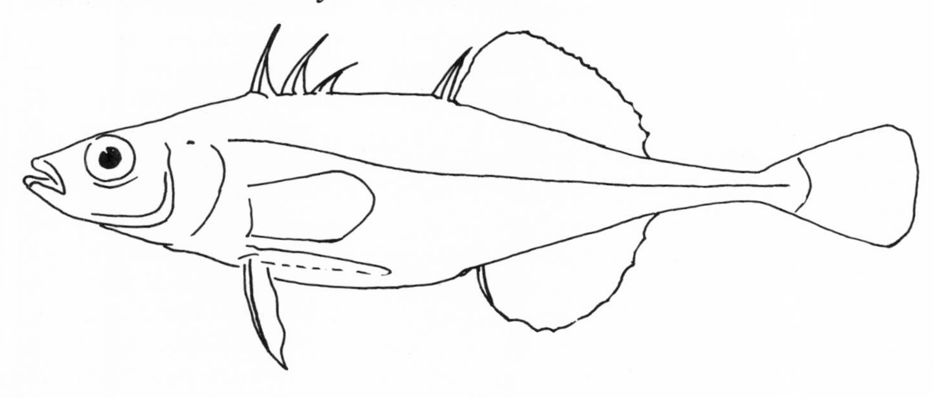

Sticklebacks are a marine, brackish, and freshwater group of seven species, of which two occur in the mid-Atlantic region. They are typically elongate and small, with a maximum total length of about seven inches, and are widely distributed in temperate and subarctic waters in North America, Europe, and Asia. Their name is derived from a series of 3 to 16 well-developed and isolated dorsal spines. The pelvic and anal fins are also preceded by a spine. Large bony scales are present on the side, or the body is naked.

The family is well known as a result of the numerous behavioral and physiological studies made of some of the species in the wild, as well as in captivity, and some studies of mating and nest-building behaviors are now classics in the animal behavior literature. Their small size, attractive shapes and colors, easy availability, and high tolerance to environmental conditions make them interesting aquarium fishes.

Fourspine stickleback

Apeltes quadracus Pl. 127

Description. 2.0 to 2.5 inches (52 to 64 mm). The fourspine stickleback typically has four (with a range of three to five) dorsal fin spines, which are of various lengths and angled alternately right and left, with the first two longer than the last two and with a wide gap present before the last. The caudal peduncle is long and very slender and the tail rounded; the skin is naked. The back is olive brown, the side a mottled deep brown, and the venter silver-white. A large male is black, and when he is reproductive, the pelvic fins are red. The **threespine stickleback**, *Gasterosteus aculeatus*, also occurs in the mid-Atlantic region but has only three dorsal spines, is more robust, and is estuarine or marine.

Distribution and Abundance. The fourspine stickleback is restricted to coastal areas from near New Bern, North Carolina, north through the Chesapeake and Delaware bays. It is more common north of the mid-Atlantic region, where it occurs to Newfoundland, Canada. It is mostly a nearshore marine and lower estuarine species, but there are numerous freshwater populations, usually in ponds and lakes. It is often common locally.

Habitat. This fish is characteristic of calm, shallow, and heavily vegetated waters.

Fourspine stickleback

Natural History. Food of the fourspine stickleback is plankton, which it pipettes out of the water; small attached or crawling invertebrates are undoubtedly eaten also. It spawns in spring and early summer. The male and female then migrate to spawning areas in shallow water, where the male establishes a territory and begins to construct a nest. The nest of aquatic plants and twigs is bound together in a complicated manner by a threadlike material secreted by the kidneys that hardens on contact with water. In an attempt to lure a female onto the partially completed nest, which at that stage resembles a cup-shaped basket, the male performs a complex mating dance in front of her. One of his conspicuous signals is a display of both red pelvic fins. If courting is successful, the male and female spawn on the partially completed nest.

The amber-colored eggs are deposited in the nest in adherent clusters of about 20 to 50, and the nest is then covered. The eggs and young are guarded by the male. He may establish one to three other nests on top of the first. The male aerates the eggs by drawing water through the nest with his mouth. Hatchlings grow rapidly, and the great majority do not live past the calendar year of life in which they were hatched.

Sculpins *Family Cottidae*

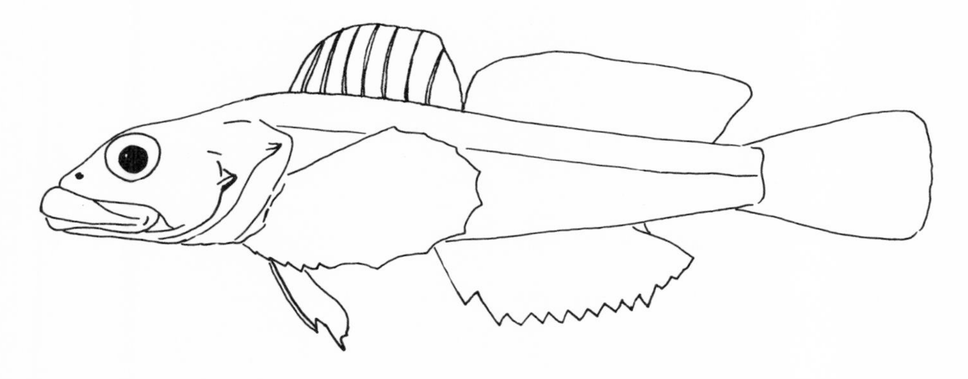

Sculpins are large-headed fishes that are widest at the head and in which the body tapers to a slender and compressed caudal peduncle. All have a large mouth, big eyes, two dorsal fins, large fan-shaped pectoral fins, a long, low anal fin, and pelvic fins that are attached to the body in the chest area. Some sculpins are naked, and others appear so because they have but few scales. Many are well protected by spines on the fins and by pointed bony processes on the skull bones. Earth-tone colors of black, brown, olive, and beige camouflage these largely bottom dwellers. Most species are small and range from about 1¾ inches to 10 inches long, but the largest attains a length of about 31 inches. They feed on animal prey living on or near the bottom. The approximately 300 species are primarily marine, but there are many estuarine and freshwater species. The group is found nearly throughout North America and the rest of the Northern Hemisphere, including Japan, and two species are known from New Zealand. In the mid-Atlantic region there are several marine species, as well as eight freshwater species.

Black sculpin
Cottus baileyi
Mottled sculpin
Cottus bairdi Pl. 128
Banded sculpin
Cottus carolinae Pl. 129
Slimy sculpin
Cottus cognatus
Potomac sculpin
Cottus girardi
Broadband sculpin
Cottus new species (includes 3 species)

Description. 2.0 to 7.3 inches (51 to 185 mm). These eight small species are all similar, and even experts sometimes have trouble distinguishing them. Their body shape is as described above for the family. The two dorsal fins run almost the length of the back, the anal fin is long, and the body is mottled brown with several dark brown to blackish bars on the back and side. The largest species of the eight reaches a maximum length of 7¼ inches, but the usual length of adults is from two to four inches.

Distribution and Abundance. The eight species have a limited distribution in the mid-Atlantic region: the black sculpin occurs in only a portion of western Virginia, the banded sculpin is limited to far western Virginia and North Carolina, and the three species currently lumped as broadband sculpin are endemic to either the Holston, Clinch, or Bluestone rivers and their tributaries in western Virginia. Of the three more widely distributed species, the mottled sculpin occurs in the mountains of the Carolinas, Virginia, and Maryland, as well as in the piedmont of northern Virginia and the coastal plain of Maryland, with isolated populations on the central Eastern Shore of Maryland (Caroline County) and adjacent southern Delaware (Sussex County). The slimy sculpin occurs in the northern half of Virginia and in several locations in Maryland and West Virginia. These populations of slimy sculpin may actually be an undescribed species. The Potomac sculpin occurs in the mountains of western Virginia, in the Potomac and Shenandoah rivers, and in tributaries of the Potomac River in western Maryland. All can be common, although the slimy sculpin is considered to be threatened in Maryland and the banded sculpin threatened in North Carolina.

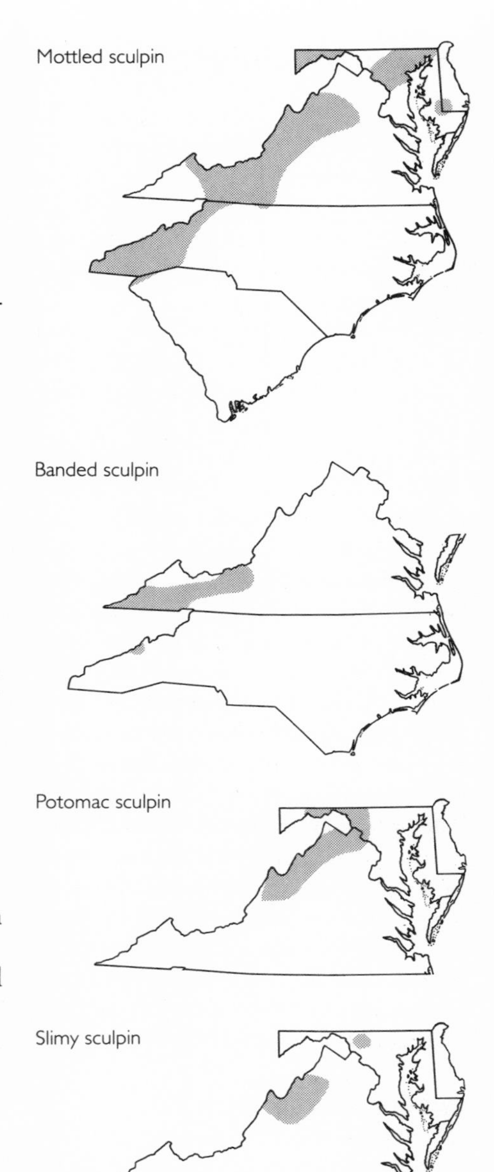

Habitat. All of the sculpins in the region are typical of clean, clear streams with a pronounced current and a rock or gravel substrate. All are restricted to environments with well-oxygenated, clear, and cool waters, with an abundance of shelter in the form of rocks.

Natural History. Sculpins spawn during the months from January to May, depending on the species. The male establishes a territory around a rock or log. The eggs are deposited in a layered clump, usually on the underside of a rock or waterlogged wood or else under or in debris such as bottles or cans, and are guarded by the male. The young grow rapidly, and some can become reproductive the year after they hatch. Only a few survive to reproduce in the year that follows their year of first reproduction. Juveniles are most common in areas of reduced current. Food consists of aquatic insects, crustaceans, small fishes, and some vegetation. Sculpins are eaten by predatory fishes and no doubt by snakes.

Temperate Basses *Family Moronidae*

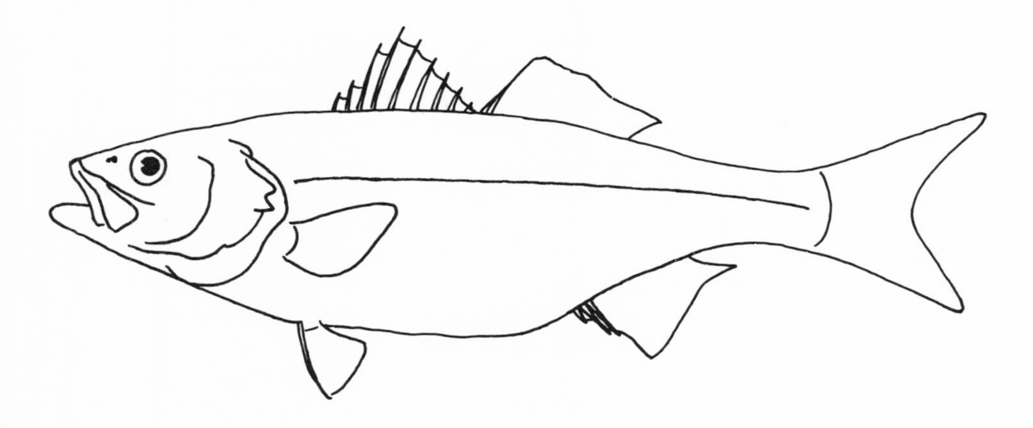

Temperate basses have a compressed and moderate or deep body, a large mouth, ctenoid scales, thoracic pelvic fins, and a complete lateral line. The anterior (spiny) and the posterior (soft rayed) dorsal fins are completely but barely separated, rather than connected as in the sunfishes. The first dorsal fin usually has nine spines, and the second dorsal fin one spine and 11 to 14 rays. There are three spines in the anal fin. The composition of this family has not been completely determined, and thus the number of species in it is uncertain.

White perch

Morone americana Pl. 130

Description. 3.9 to 19.0 inches (100 to 483 mm). The white perch has an olive green back and a silvery side and venter. The adult lacks the distinct dark lateral stripes of the white and striped basses.

Distribution and Abundance. This species is found in coastal waters from Canada to South Carolina. It is an anadromous species that has naturally become landlocked in some lakes. It has been introduced into some lakes and reservoirs. It may be locally abundant.

Habitat. It is found in brackish waters and some lakes, such as Lake Mattamuskeet in North Carolina. It has frequently been stocked.

Natural History. The white perch is an open water schooling fish. It is anadromous and migrates up coastal streams or tributaries of lakes and reservoirs to spawn in spring or early summer. Females simultaneously release their eggs, which are then fertilized by nearby males. The eggs adhere to tree branches and other firm substrates and receive no parental care. Upon hatching, the fry are fully independent and begin to work their way downstream to a lake or estuary. Diet changes with size, and food includes insects and other invertebrates as well as small fishes. The male is mature at two years of age, the female at four. The maximum age is 17 years, and few individuals live beyond age seven. It is usually a small fish, averaging less than one-half pound in lakes and reservoirs. In estuarine environments it becomes larger, and the maximum weight in the

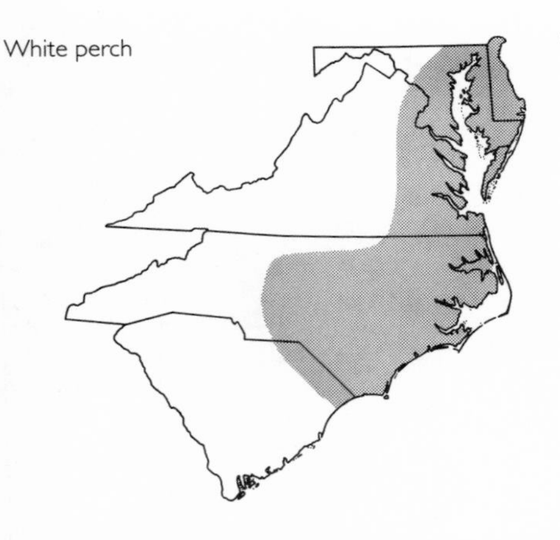

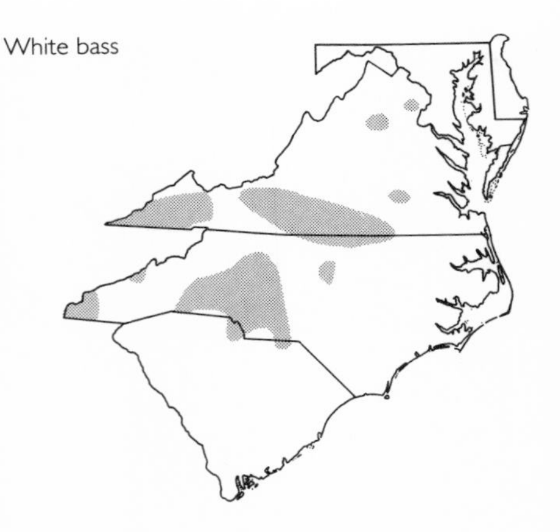

mid-Atlantic region is almost two pounds. It provides good fishing when a school can be located and readily takes a variety of natural and artificial bait.

White bass

Morone chrysops Pl. 131

Description. 10.8 to 17.7 inches (275 to 450 mm). The body is silvery with seven solid or broken narrow stripes on the side. The body is deeper than that of its similar-looking relative, the striped bass. The maximum weight is about six pounds.

Distribution and Abundance. Native to the Mississippi River basin, this species has been stocked in reservoirs in the Carolinas and Virginia. It is locally abundant.

Habitat. The white bass prefers cool, clear, and open waters, where it often travels in schools and follows forage fishes. It may occur both in shallow and deep portions of lakes, depending on water temperature and food availability. In the mid-Atlantic region it is most successful in large reservoirs in the piedmont and mountains.

Natural History. The white bass begins to spawn in April or May when the water temperature reaches 60°. It then migrates from open waters into tributary streams, sometimes for a distance of 40 miles. It constructs no nest, and spawning is similar to that of the white perch and striped bass. Up to a million eggs are released and fertilized near the water surface, and these settle to the bottom and stick to the substrate. Growth of the fry is rapid and maturity can be attained in two years, although three years is more typical; it can live for at least 10 years. This schooling species feeds primarily on young threadfin and gizzard shads. White bass and striped bass have been hybridized in fish hatcheries, and the appearance of the hybrid is intermediate between that of the two parents. The hybrid grows larger than the white bass, but its habitat requirements and behavior are similar to that of the white bass. Both the white bass and the white bass–striped bass hybrid are excellent game fishes.

Striped bass

Morone saxatilis Pl. 132

Description. 15.7 to 78.7 inches (400 to 2,000 mm). This robust, elongate fish, often called rockfish or rock, has an olive green back, a silvery side with a series of thin dark stripes, and a silvery white belly. It differs from the similar-appearing white bass by its more elongate body, larger size, and more numerous lateral stripes.

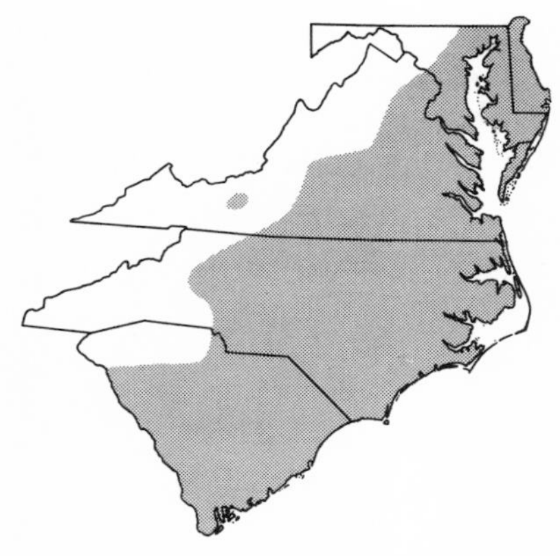

Striped bass

Distribution and Abundance. The striped bass is native to estuarine coastal rivers of eastern North America and to the adjacent ocean. It occurs from Canada to the Gulf of Mexico. Once an abundant and important commercial and sport fish, it has suffered a serious decline over most of its range in recent years. It has been successfully introduced into many freshwater impoundments in and outside of the mid-Atlantic region, and on the West Coast of the United States in the 1860s.

Habitat. In the ocean it is found from breaking waves on the beach to several miles offshore over sand and mud substrates. It is often caught around piers and jetties as it migrates and/or as it seeks food. In estuarine areas it occurs over mud and sand bottoms, near marsh vegetation and rock jetties, and in rivers with shelter of submerged tree trunks and branches. While it survives well in some large reservoirs where it has been stocked, it reproduces only in those few reservoirs that provide suitable and specialized conditions.

Natural History. The anadromous striped bass spawns in early spring after migrating up coastal rivers or up tributary rivers in the case of landlocked individuals. One female may lay a million eggs. The eggs are slightly heavier than water and are suspended in the water column and "bounce" along the bottom as they drift downstream toward a coastal sound or impoundment. They receive no parental care. Should they reach nonflowing water prior to hatching, the eggs will settle to the bottom and die. Thus, an important characteristic of a good spawning river is enough distance from the spawning area to a sound or lake for the eggs to develop and hatch before they reach still water. The fry feed on plankton and on larger prey when older. The first year for non-landlocked fish is spent in coastal estuaries, after which they migrate to the ocean. The female will return to spawn for the first time when four or five years old. The ability of the striped bass to survive in only fresh water was discovered when searun striped bass were cut off from the ocean as a result of the construction of the Santee-Cooper Reservoir on the coastal plain of South Carolina. This population maintained itself and then increased in numbers as adults moved out of the reservoir and upriver to spawn. The juvenile of this population lives in the open reservoir waters.

The adult striped bass is a voracious feeder on other fishes, especially schooling species such as menhaden, herrings, and gizzard and threadfin shads, as well as the American eel. It reaches a weight of over 100 pounds and is valuable to both commercial and sport fishermen. A recent decline in abundance has resulted in severe restrictions by the authorities in Atlantic coast states on the method of taking striped bass and on the size and quantity of individuals that may be caught. Much research is underway to understand the reasons for the decline and for ways to foster recovery.

Sunfishes *Family Centrarchidae*

The 30 species in this family include the sunfishes, crappies, and black basses. All have a moderately deep to very deep body, and the anterior spiny and posterior soft portions of the dorsal fin are joined. They resemble the temperate bass (Moronidae) and sea bass (Serranidae) families. Sunfishes are native to the warmer waters of North America east of the Rocky Mountains. They have been widely introduced throughout the United States, however, and some, especially the largemouth bass, have been stocked widely throughout the world.

Mud sunfish

Acantharchus pomotis Pl. 133

Description. 5.6 to 8.6 inches (142 to 218 mm). This sunfish is stocky and oval and has a rounded caudal fin. The snout is short, the mouth and eye large, and the overall color olive green with three to six dark brown stripes on the side. It is the only sunfish that has cycloid scales (scales that lack backward-pointing and microscopic teeth and that feel smooth when the fingertips are run along the fish from the tail to the head).

Distribution and Abundance. This is primarily a fish of the coastal plain, and it is found along the Atlantic Coast from southern New York to northern Florida. It is widely distributed throughout the mid-Atlantic region but is rarely found in numbers: few fishermen have ever seen it, even though many fish the waters in which it occurs.

Habitat. This fish is found in lowland swamps, pools, sloughs, and backwaters of creeks where submerged vegetation is abundant and the bottom is of silt,

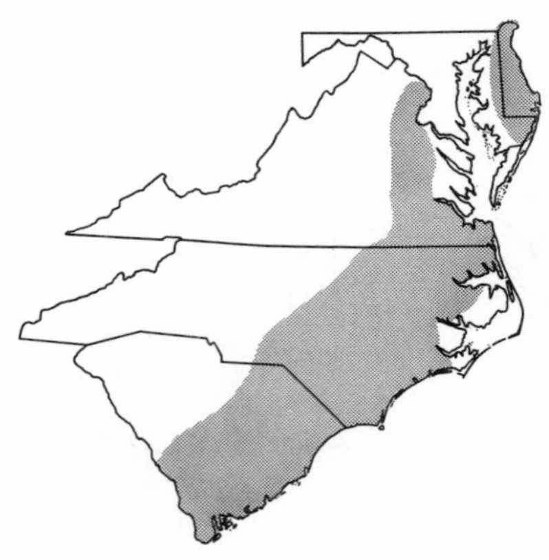

Mud sunfish

mud, and detritus. These waters are usually acidic and tea-colored. The combination of a muddy bottom and heavy vegetation make collecting it difficult.

Natural History. The mud sunfish is nocturnal and by day hides under banks or in dense vegetation, sometimes partly buried in mud. It often rests head down among plants. In North Carolina it spawns from December to May with low but rising water temperatures. It builds a saucer-shaped nest in sand surrounded by mud and aquatic vegetation. A female contains up to 3,800 eggs. Growth is rapid, and maturity is reached in the second year. Maximum age attained in North Carolina is four years, while ages of up to seven years were found in mud sunfish from Maryland. The food is scuds, crustaceans, and larvae of aquatic insects. This fish reportedly can make a grunting sound, something unique among the sunfishes.

Roanoke bass
Ambloplites cavifrons
Rock bass
Ambloplites rupestris Pl. 134

Description. 5.7 to 16.9 inches (144 to 430 mm). Often called redeye bass or goggle eye, these two similar-appearing species have a short, robust body with an overall dark olive brown color and a red eye. Their side is covered with dark spots that tend to form rows. They have five or six anal fin spines, while most other sunfishes have three. The Roanoke bass, with its unscaled or partially scaled cheek and numerous iridescent gold to white spots on the head and upper body, differs from the similar rock bass, with its scaled cheek and head without spots. These two basses have hybridized in the upper Roanoke River.

Distribution and Abundance. The Roanoke bass occurs in only four river drainages in North Carolina and Virginia: the Chowan, Roanoke, Tar, and Neuse. It is uncommon and, more often, rare. The rock bass in the mid-Atlantic region is widely distributed and common west of a diagonal line drawn from extreme western South Carolina to the head of Chesapeake Bay. It is also common in much of the northeastern quarter of the United States and Canada. It has been widely and successfully introduced into Atlantic slope drainages, including, unfortunately, the upper Roanoke River, where it has supplanted the Roanoke bass.

Habitat. Both species inhabit cool and warm creeks and rivers that have a rock and gravel bottom. Most individuals are caught in deeper water around shelter such as large boulders, tree roots, or branches. The rock bass is uncommon in lakes and reservoirs, and the Roanoke bass is rare there.

Natural History. Both species have a similar biology. They reach maturity when two years old and from three to

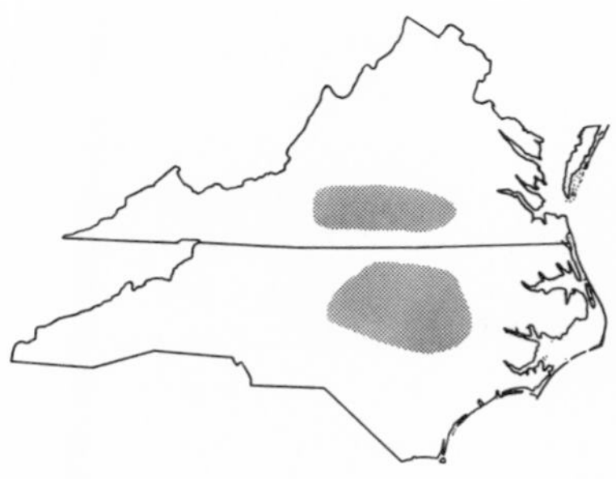
Roanoke bass

Rock bass

five inches long. In May and June the male makes a saucer-shaped nest depression in sand or gravel, as is typical of sunfishes. The female visits the nest only to lay her adhesive eggs, which total 3,000 to 11,000, and the male guards them until the fry leave. After one year of life the young average slightly more than three inches in length. The diet is crayfishes, aquatic insects, and fishes. Both species provide excellent fishing with light spinning tackle and fly rods and will take a variety of bait and lures. The maximum age of the rock bass is approximately 10 to 12 years, and 9 for the Roanoke bass.

Flier

Centrarchus macropterus Pl. 135

Description. 2.8 to 7.5 inches (70 to 190 mm). The flier is of medium length and is deep bodied; the background color is olive green to silver, there are several rows of dotted linelike brown spots on the side, and the fins are heavily marked with light spots. A black teardrop-shaped bar usually occurs below the eye, and there is a black spot bordered in orange on the rear of the dorsal fin in the young, which fades and disappears as the fish grows.

Distribution and Abundance. This warmwater sunfish occurs on the coastal plain from southern Maryland to Florida and on the Gulf slope into the lower Mississippi River basin. It is often abundant.

Habitat. The flier inhabits warm, heavily vegetated waters such as roadside drainage ditches, sloughs, ponds, and calm portions of streams.

Natural History. The flier spawns in early spring when the male builds a nest and defends a small territory, and this

Flier

and other reproductive behavior is probably similar to that of other sunfishes (see the bluegill account below). The female lays up to 36,000 eggs, and the nest and the newly hatched fry are guarded by the male. The flier feeds largely on insects and crustaceans, but larger individuals also eat small fishes. Maturity may be reached at the end of the first year when it is 2½ to 3 inches long, and the maximum age reached is approximately seven years. This small fish is sometimes taken by fishermen seeking larger sunfishes. It takes a range of natural and artificial bait. It makes an attractive aquarium fish, especially when juvenile.

Blackbanded sunfish

Enneacanthus chaetodon Pl. 136

Bluespotted sunfish

Enneacanthus gloriosus Pl. 137

Banded sunfish

Enneacanthus obesus Pl. 138

Description. These three fishes can be separated from all other sunfishes by their small size as adults and by the combination of three anal fin spines and a rounded caudal fin. Other sunfishes with three anal spines have a forked caudal fin.

Blackbanded sunfish: 2.0 to 3.2 inches (50 to 82 mm). This is the most distinctive of the three species. Both the juvenile and the adults of both sexes have six black bars on a silvery body and a leading edge of the dorsal and pelvic fins that is distinctly red or pink.

Bluespotted sunfish: 2.0 to 3.8 inches (50 to 95 mm). This sunfish is profusely covered with iridescent blue, green, silver, or gold spots. These spots are most developed on the reproductive male, in which they occur on an almost black body. The young and the nonreproductive adults usually have weakly defined dark bars on the side.

Banded sunfish: 2.0 to 3.8 inches (50 to 95 mm). This species is similar to the bluespotted sunfish, but it is much less conspicuously spotted and there are several uniform dark bars on the side, which are most conspicuous in the reproductive adults.

Distribution and Abundance. All three species are widely distributed on the coastal plain in most Atlantic and eastern Gulf slope drainages. The distribution of the blackbanded sunfish is disjunct compared to that of the other two species, and it is unexpectedly absent from some drainages where apparently appropriate habitat exists. It is sometimes locally common but is usually uncommon. It is considered to be endangered in Virginia. Both the bluespotted and the banded sunfish are usually common.

Habitat. All three species are found in calm, vegetated waters of lakes, ponds, pools, roadside ditches, creeks, and small rivers over a sand or mud bottom. The blackbanded sunfish appears to be restricted to more heavily vegetated, darkly stained, and highly acidic waters than its two relatives.

Natural History. These sunfishes feed mostly on aquatic insect larvae, scuds, and other small crustaceans. Spawning can be prolonged and may occur from early spring into summer. The male constructs a small circular nest, and the adhesive eggs are deposited there. Subsequent spawning may occur in a nest after earlier eggs have hatched. Growth is rapid the first year and slows substantially thereafter. The maximum age is four years. All three species adapt readily to aquarium conditions, and the black-

Blackbanded sunfish

Bluespotted sunfish

Banded sunfish

banded sunfish has been particularly popular with European and Asian aquarists.

Redbreast sunfish

Lepomis auritus Pl. 139

Description. 2.4 to 9.4 inches (60 to 240 mm). This sunfish, called robin in many areas, is more elongate than most others. The throat and belly are bright orange in the male and yellow in the female. The upper body is usually uniformly greenish but in the breeding male has a sky blue sheen. The female has the same general colors but is more dull. The opercular lobe is black and long. The pectoral fin is pointed. The gill rakers are short and hard.

Distribution and Abundance. This species occurs throughout the mid-Atlantic region except in cold mountain waters. It is native to drainages on the Atlantic slope from New England to Florida, and it has been transplanted to other areas in recent years. It is usually common.

Habitat. The redbreast sunfish is found primarily in running water in streams from mountain foothills to the coast.

Natural History. It spawns from April to June on a sand bottom, often after an upstream migration. It builds a nest similar to that of the bluegill, but nests are in smaller groups and situated farther apart. The nest is near shelter, such as stumps and logs, and is often one that has been abandoned by other sunfishes. A female can lay up to 14,000 eggs. The redbreast sunfish reaches sexual maturity in three to four years, when as short as four inches; the maximum age attained is approximately nine years. Food includes aquatic insects, small clams,

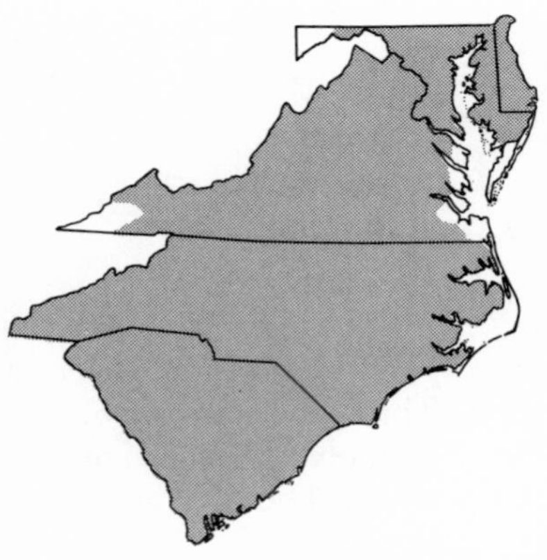
Redbreast sunfish

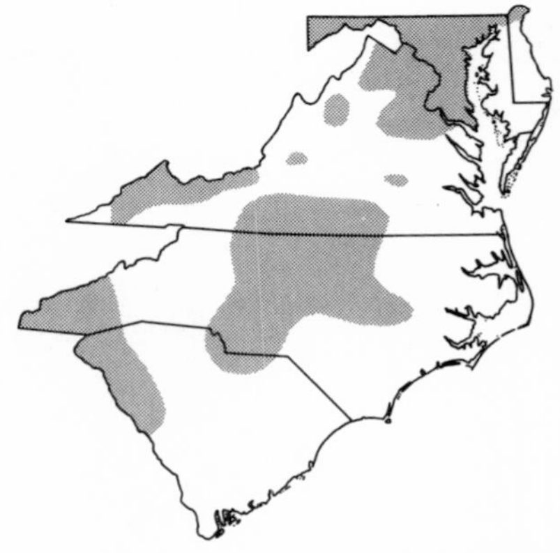
Green sunfish

crustaceans such as crayfishes, and small fishes. This fish seldom weighs more than one pound, although a record one pound, eight ounce, redbreast sunfish was caught in Florida. It apparently reaches its largest size in blackwater streams of the coastal plain.

Green sunfish

Lepomis cyanellus Pl. 140

Description. 5.9 to 12.2 inches (150 to 310 mm). This fish is more elongate and basslike than other sunfishes. It is green above and has a yellow or orange belly. The mouth is large, the pectoral fin short and rounded, and the opercular lobe dark, but with a pale edge, and stiff (not flexible at the end).

Distribution and Abundance. Within the mid-Atlantic region, the green sunfish is native only to extreme western North Carolina, Virginia, and Maryland, in tributaries of the Mississippi River. It is also native west of the region, from Canada to the Gulf of Mexico. Through introductions it is now found in much of the mid-Atlantic region, and elsewhere, although it is absent from most of the region's lower coastal plain. It is usually uncommon but may be locally common.

Habitat. It is primarily a stream dweller, usually in small and rocky piedmont and mountain foothill creeks.

Natural History. It spawns in spring and summer. The nest is like those of other sunfishes (see the bluegill account below) but is either isolated or occurs in only small groups, and it is usually located in pools or along stream banks. The male defends it. The male makes a grunting sound as he courts the female. Maturity is sometimes reached in one year, but it usually requires two, and the maximum age attained is 10 years. In confined areas, individuals may be stunted. Food includes aquatic insects such as larvae of dragonflies and caddisflies, terrestrial insects that fall into the water, and small fishes. It is a pioneer species and readily invades new bodies of water. The green sunfish can tolerate greater turbidity than many other sunfishes. Although not sought by fishermen, it readily takes bait.

Pumpkinseed

Lepomis gibbosus Pl. 141

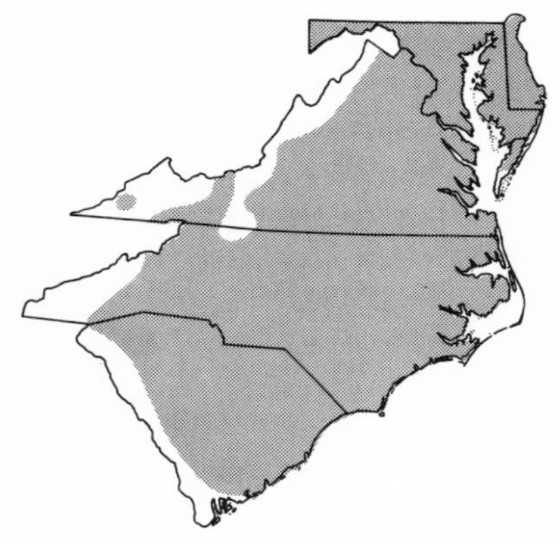

Pumpkinseed

Description. 3.1 to 15.0 inches (78 to 381 mm). The pumpkinseed is identified by a greenish upper body and a side that is heavily speckled with black and orange, which looks like seeds in a pumpkin. The cheek is striped with blue and orange, and the throat and belly are bright orange; these are brightest in the breeding male. The opercular lobe is short, and its posterior edge is orange bordered with white (which is an important species characteristic).

Distribution and Abundance. This fish is found throughout the mid-Atlantic region except western Virginia, western North Carolina, and a portion of South Carolina that parallels the Savannah River. It is native to most of the northeastern quarter of the United States and adjacent Canada, and it has been widely stocked elsewhere. It is often abundant.

Habitat. Although often found in cooler water than other sunfishes, it thrives in farm ponds, old mill ponds, and quiet coves of reservoirs, especially where there is heavy vegetation.

Natural History. The biology of the pumpkinseed is similar to that of our other sunfishes. It spawns in spring and early summer, on a nest built in shallow water, usually near the edges of ponds and lakes, and vigorously defended by the male. The male may spawn in the same nest with two females. A female may lay up to 3,000 eggs. After they hatch, the young remain near the shallow-water breeding area. Most pumpkinseeds mature at two years of age; the maximum age attained is 12 years, but few individuals exceed six years. The pumpkinseed eats a variety of food such as insects, snails, clams, and small fishes, grows slowly, and seldom exceeds a length of five inches. It is active in the day and rests on the bottom at night. Anglers who seek the larger sunfishes often catch it with crickets, worms, popping bugs, and small spinners. It is an excellent species for the beginner with only a pole and a worm.

Warmouth

Lepomis gulosus Pl. 142

Description. 3.0 to 12.2 inches (75 to 310 mm). The warmouth is a mottled olive green, often with a yellow or orange wash on the belly. Three dark lines extend from the eye onto the operculum, the opercular lobe is short and black with a light border, the mouth is large, and the maxillary bone extends back past the middle of the eye. The body is more elongate than in most sunfishes.

Distribution and Abundance. In the mid-Atlantic region the warmouth occurs in both the Carolinas, Virginia, the southern half of Maryland, and Delaware (an isolated record). It is found from the mountains to the coast but is most abundant in the piedmont and coastal plain. It also occurs through

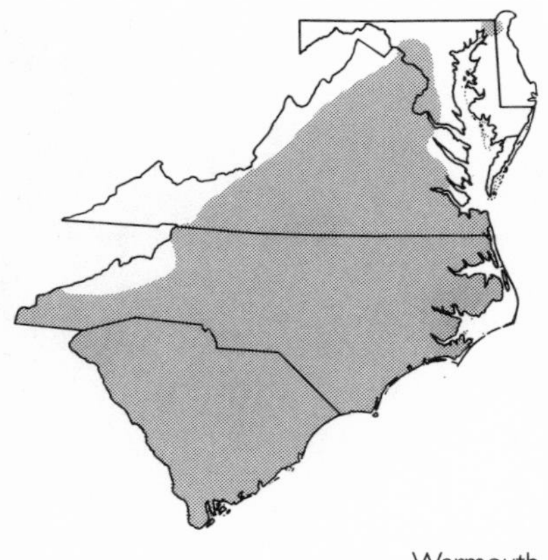

Warmouth

much of the eastern United States from the southern Great Lakes region to the Gulf of Mexico. It is usually uncommon.

Habitat. The warmouth prefers slow streams, ponds, and lakes with extensive submerged vegetation and a mud or detritus bottom, and it is common among water lilies in old mill ponds and slow coastal rivers. It is reported in brackish water where the salinity can reach four parts per thousand.

Natural History. It breeds from April to August, and the male builds a nest as do other sunfishes. The nest is often solitary and placed next to a stump or underwater log on a bottom of loose silt or rubble covered with silt. The male may spawn with a number of females, each of whom can lay up to 63,000 eggs. He guards the nest until the eggs hatch and the fry depart, which usually occurs five or six days after spawning. Some fish mature at one year of age, but most require two, and the maximum age reached is eight years. The warmouth is more piscivorous than other sunfishes, and it also eats crayfishes and insects. It is considered an excellent small game fish and will take a variety of bait.

Bluegill

Lepomis macrochirus Pl. 143

Description. 7.0 to 16.1 inches (178 to 410 mm). The bluegill has an oval shape and a small mouth. The upper back is blue or purple, the side usually has black bars, there is a dark spot at the base of the posterior end of the dorsal fin, the opercular lobe is black, and the throat is orange. The juvenile is often pale but with distinct bars on the side, a dark opercular lobe, and a dark spot on the dorsal fin.

Distribution and Abundance. This species is native to much of the eastern and central United States and adjacent Canada and Mexico. It has been widely stocked and now occurs throughout the mid-Atlantic region and almost all of the United States and northern Mexico, as well as in Europe and South America. It is oftén abundant.

Habitat. The bluegill inhabits warm slow or still water, usually near aquatic vegetation. It is absent from cold mountain streams and lakes. It is often stocked in farm ponds, lakes, and reservoirs.

Natural History. The bluegill spawns in spring or early summer when the water temperature reaches about 70°, and the peak of spawning in the mid-Atlantic region is in May or June. The male then moves into shallow water, usually over a bottom of sand or gravel and fans out a saucer-shaped depression one to two feet in diameter and a few inches deep. Nests are grouped and may number from less than a dozen to several hundred. The female is attracted to the site and after a brief courtship lays from a few thousand to as many as 60,000 small, adhesive eggs in one or more nests. She may spawn several times in

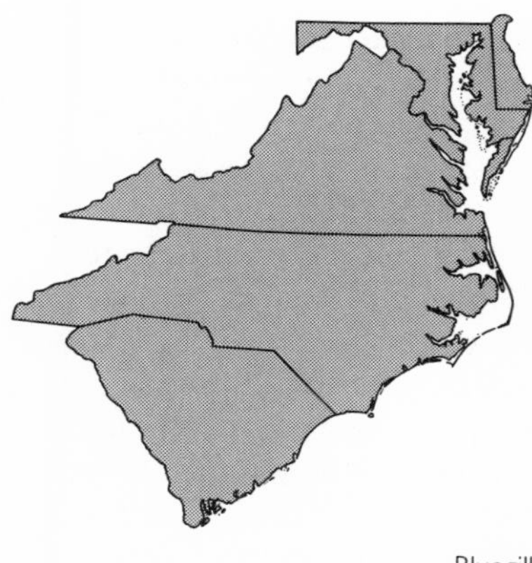

Bluegill

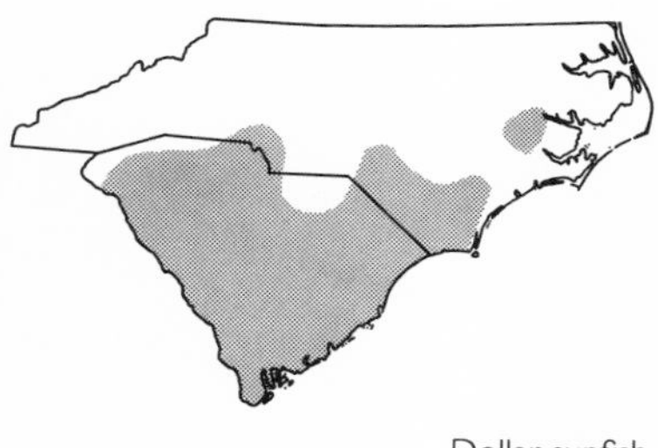

Dollar sunfish

one season. The male fans the eggs to aerate them and to keep them free of debris. He also guards the nest until the fry hatch, usually in two to five days, and then guards the young until they leave. Some bluegill spawn at one year of age, but most are not ready until age two; the maximum age reached is 11 years. The bluegill feeds on aquatic insects and their larvae, small fishes, and crustaceans such as crayfishes. When a population is not crowded and food is plentiful, it grows relatively rapidly; adults commonly reach a half to three-quarters of a pound, but huge individuals of up to 4 pounds, 12 ounces, have been caught. However, it can severely overpopulate small ponds when predation is low, and then its growth is slow and the maximum size often less than four ounces.

The bluegill is one of the most heavily fished species in the mid-Atlantic region. It is widespread, has a voracious appetite, and can be caught with unsophisticated gear. It readily takes both natural and artificial bait.

Dollar sunfish

Lepomis marginatus Pl. 144

Description. 1.7 to 4.7 inches (43 to 120 mm). This small sunfish seldom reaches 4½ inches. It is short and deep bodied, olive green in color, and has orange on the cheek and belly. Wavy blue lines radiate from the eye onto the operculum. The opercular flap is dark, elongate, and flexible, with a pale green margin.

Distribution and Abundance. This fish occurs in central and southeastern North Carolina and all of South Carolina except the mountains. It also occurs in much of the Atlantic and Gulf coastal plains and the lower Mississippi River Valley. It is usually uncommon.

Habitat. A fish of warm coastal waters, it is associated with dense aquatic vegetation and a mud or detritus bottom.

Natural History. The dollar sunfish spawns from May to August. It matures in its second year and can reach a maximum age of five years. It may nest in large groups, in shallow water on a sand bottom. Food is primarily aquatic insects and crustaceans.

Longear sunfish

Lepomis megalotis Pl. 145

Description. 1.9 to 9.4 inches (49 to 240 mm). The brightly colored longear sunfish appears similar to the redbreast, but

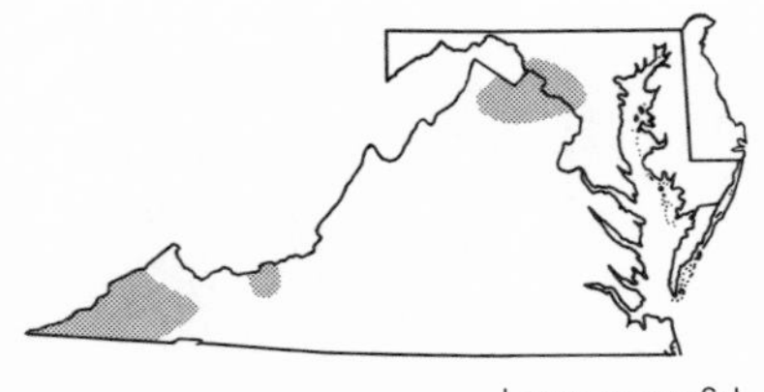
Longear sunfish

it has a deeper body and an opercular lobe wider than the eye and bordered with white, blue, or orange (the opercular lobe in the redbreast is not as wide as the eye and is black).

Distribution and Abundance. The longear sunfish is native to the Mississippi River basin, and in the mid-Atlantic region occurs naturally only in westward flowing streams in the mountains of Virginia, where it is usually uncommon. It has been stocked in reservoirs in the region.

Habitat. This sunfish occurs most often in small mountain streams and now also in reservoirs.

Natural History. The longear sunfish spawns from May to August, in a nest typical of other sunfishes. The species is gregarious and nests are often in groups. A female can lay up to 4,000 eggs. Sexual maturity is attained at an age of two years, and the maximum age is eight. Aquatic insects, other invertebrates, and small fishes comprise the bulk of its food.

Redear sunfish

Lepomis microlophus Pl. 146

Description. 5.3 to 15.0 inches (134 to 381 mm). This fish resembles a pumpkinseed, but the rear margin of the gill cover is thin and flexible, and it lacks the characteristic blue cheek stripes of the latter. It also has a pale yellow or orange spot on the margin of the opercular lobe. The redear sunfish has heavy teeth in the throat, which are used to crush snails and mussels, giving rise to a common local name, "shellcracker."

Distribution and Abundance. The redear sunfish is native to most of the southeastern one-fourth of the United States, including the central Carolinas exclusive of the mountains and foothills. It is not widespread in the mid-Atlantic region, but it can be locally abundant. It has been widely introduced elsewhere.

Habitat. It occurs in ponds, lakes, and slow streams, in clear and warm water, and is usually associated with aquatic vegetation or other underwater cover.

Natural History. The redear sunfish begins to breed at an age of two years. The nest is often built near aquatic vegetation, and nests may occur in large concentrations. It spawns throughout the warm season, and nest building, maintenance, and defense, as well as care of eggs, is similar to that of the bluegill. A female can lay up to 45,000 eggs. It forages at or near the bottom on a wide va-

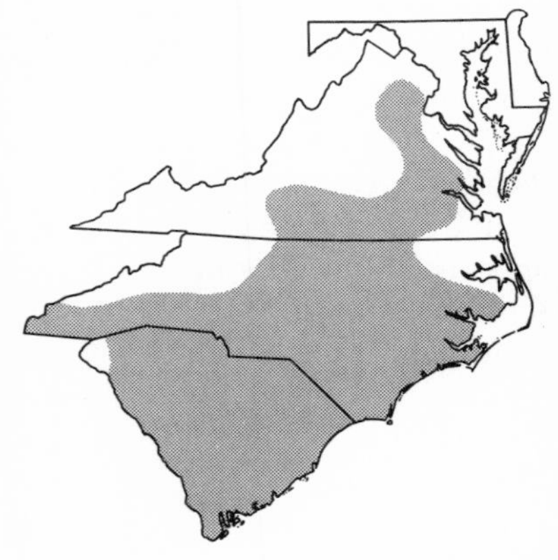
Redear sunfish

riety of food, and the heavy teeth in the throat allow it to crush snails, small clams, and other hard invertebrates. It grows larger than most sunfishes and can exceed a weight of four pounds, although just over two pounds is much more common.

Spotted sunfish

Lepomis punctatus Pl. 147

Description. 2.2 to 7.9 inches (55 to 200 mm). Often called stumpknocker, this relatively drab fish has an olivaceous body with many brown or black spots on the side and often with an orange wash on the throat and belly. The opercular lobe is short and dark, with a bony margin.

Distribution and Abundance. This southern species occurs from southeastern North Carolina south throughout most of Florida and west into the Florida panhandle. It is usually uncommon.

Habitat. It inhabits ponds and slow streams in or near dense aquatic vegetation.

Natural History. In the mid-Atlantic region the spotted sunfish spawns from early spring through summer, with the peak in May. The general pattern is similar to that of most other sunfishes, except that the spotted sunfish is a solitary nester. It matures at two years of age. This species usually feeds on the bottom, on many species of insects and aquatic algae, but it will also feed at the surface.

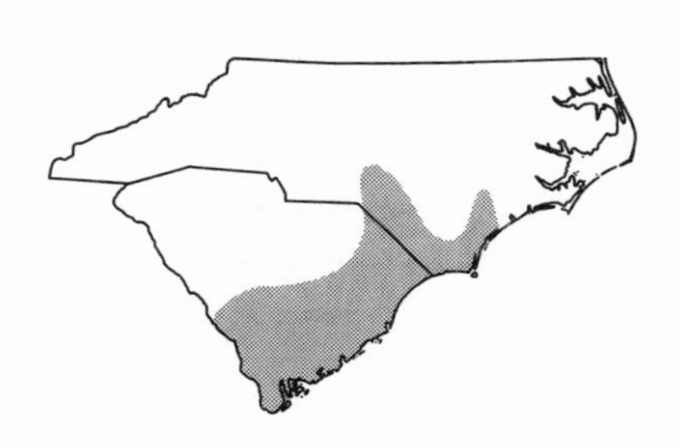

Spotted sunfish

Redeye bass

Micropterus coosae Pl. 148

Description. 5.7 to 15.0 inches (144 to 381 mm). The redeye bass resembles the smallmouth bass but is smaller. The maxillary extends to the back of the eye, while in the similar largemouth bass it extends past the eye. There is a series of dark bars on the side, each with a light center, and these may be replaced by a row of blotches in the old adult. In South Carolina the usual weight of an adult fish is between six ounces and one pound.

Distribution and Abundance. This small bass is native to only a few rivers in western South Carolina, southwestern North Carolina, and parts of northern Alabama and Georgia. In the mid-Atlantic region it is native only in the upper Savannah River drainage; it has been stocked in Lake Hartwell, South Carolina. It is usually uncommon.

Habitat. This bass appears to prefer the cool waters of piedmont or mountain foothill streams, although in Lake Hartwell it apparently grows better than in streams.

Natural History. Reproduction is similar to that of other sunfishes. The male constructs a nest, eggs are deposited there, and the nest is then guarded by the male. This bass is primarily insectivorous.

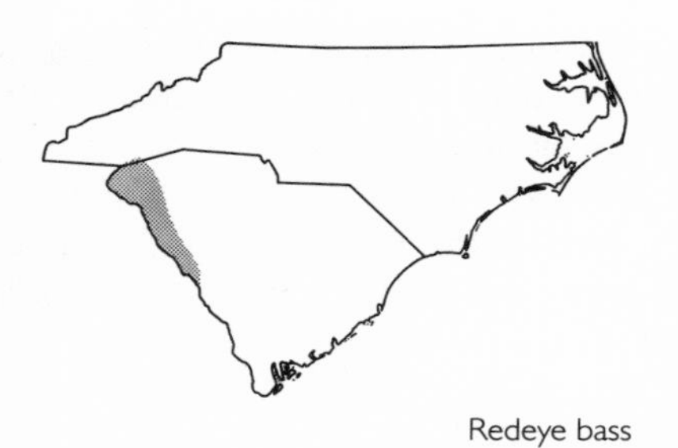

Redeye bass

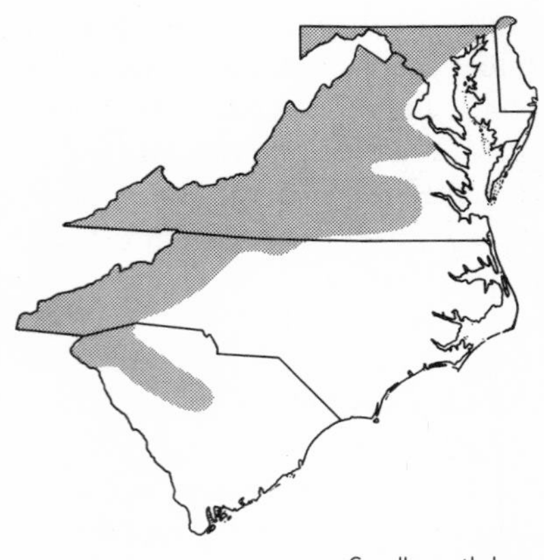

Smallmouth bass

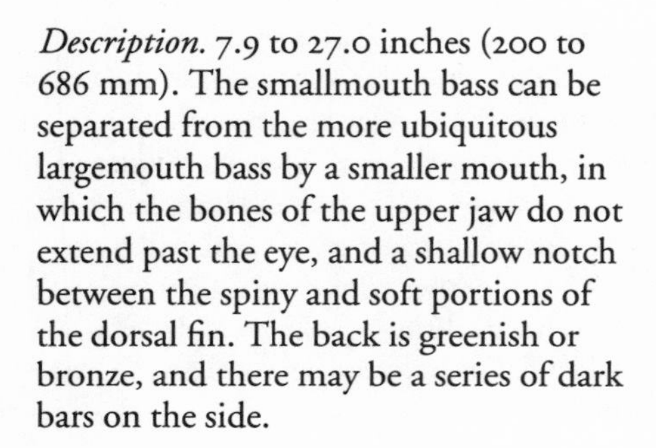

Smallmouth bass

Micropterus dolomieu Pl. 149

Description. 7.9 to 27.0 inches (200 to 686 mm). The smallmouth bass can be separated from the more ubiquitous largemouth bass by a smaller mouth, in which the bones of the upper jaw do not extend past the eye, and a shallow notch between the spiny and soft portions of the dorsal fin. The back is greenish or bronze, and there may be a series of dark bars on the side.

Distribution and Abundance. This bass is native to the upper and middle Mississippi River basin, and it has been widely introduced throughout North America and elsewhere. It now occurs through the mountainous portions of the mid-Atlantic region and extends onto the piedmont and even the coastal plain in appropriate habitat. Like most other predators, it is usually uncommon.

Habitat. The smallmouth bass is restricted to clear waters, though they can be either cool or warm, usually mountain rivers and deep reservoirs. It is intolerant of turbidity and siltation.

Natural History. It spawns from late spring to early summer. A male fans out a nest two to four feet in diameter in a gravel bottom, usually in flowing water at a depth of one to four feet. He may spawn with several females. A female can lay up to 21,000 eggs. The solitary nest is guarded by the male until the eggs hatch and the fry depart. Most individuals spawn for the first time when three or four years old. Most adults are between three and seven years of age, and the maximum age reached is 12 years. It feeds on insects when young and later feeds primarily on crayfishes and fishes. This is an important game fish in our mountains, and while it does not grow as large as the largemouth bass, it is a real challenge to catch. The habitat where it occurs is usually pristine, and, in fact, the presence of a large population of smallmouth bass is usually an indicator of a healthy stream.

Spotted bass

Micropterus punctulatus Pl. 150

Description. 11.8 to 24.0 inches (300 to 610 mm). This fish is similar to the more common smallmouth bass. The back is greenish and the abdomen whitish, on the side there is a row of connected dark blotches that usually ends in a distinct black blotch just before the tail, and there is a shallow notch between the spiny and soft portions of the dorsal fin. The scales are usually spotted, the bones of the mouth do not extend back past the eye (they do in the largemouth bass), and there are usually 12 dorsal rays (rather than the 13 to 15 of the smallmouth bass).

Distribution and Abundance. It is native to the Mississippi River basin and much of the Gulf slope from central Texas through the Florida panhandle. In the mid-Atlantic region it occurs naturally in western North Carolina and western Virginia. It has been introduced into the upper Cape Fear River drainage in North Carolina and into the Roanoke, James, and York drainages in Virginia. It can be common.

Habitat. The spotted bass inhabits cool and warm large mountain streams and reservoirs, and in the former it is usually found in deep pools over gravel. It tolerates warm water and turbidity better than the smallmouth bass, and it sometimes thrives where the latter is in decline.

Natural History. The reproductive biology of this species is similar to that of the smallmouth bass. It spawns in late spring or early summer, and the saucer-shaped nest is usually solitary and guarded by the male until the eggs hatch and the fry are large enough to leave, ordinarily in eight to ten days. Food includes insects, crustaceans, and fishes. Maturity is reached at two to three years of age, and few individuals live beyond six years. In portions of the Mississippi River basin it is an important sport fish.

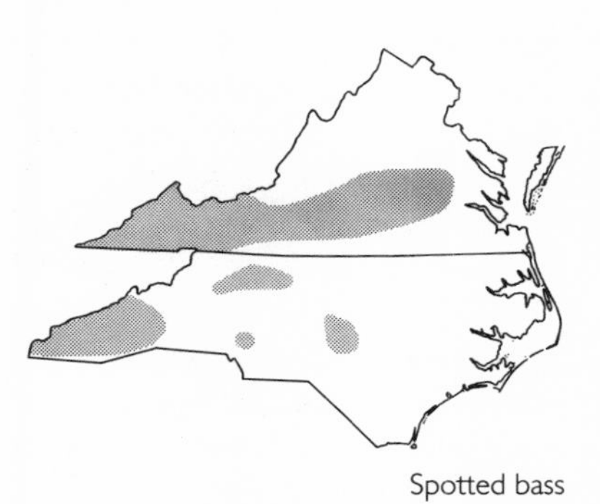

Spotted bass

Largemouth bass

Micropterus salmoides Pl. 151

Description. 4.7 to 38.2 inches (120 to 970 mm). The largemouth bass is the largest sunfish. It has a moderately elongate and slightly compressed body, a large mouth, and an upper jaw that extends beyond the rear margin of the eye. The spiny and soft portions of the dorsal fin are nearly separated. The dorsum is greenish and grades to white on the belly. A series of dark blotches form a ragged dark stripe along the side from the head to tail. This stripe becomes indistinct in larger individuals.

Distribution and Abundance. In the mid-Atlantic area, the largemouth bass originally occurred only in South Carolina. It has, however, been widely introduced and now occurs throughout the region, as well as throughout most of the United States and in many other countries. It is often a highly successful fish and is locally abundant.

Habitat. This bass prefers warm, calm, clear water. It thrives in slow streams, farm ponds, and lakes and reservoirs with high water quality. It is often replaced by the smallmouth bass in cooler mountain lakes and streams.

Natural History. In spring a male defends a nest and the area around it, a territory some seven to ten feet in diameter. The nest is a circular area of clean

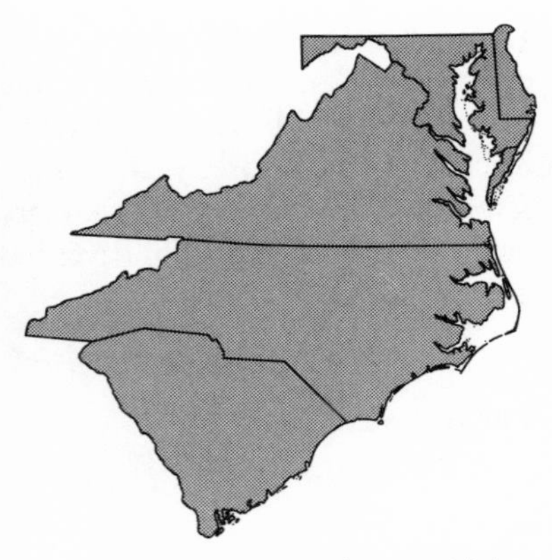

Largemouth bass

sand from which all organic debris and silt have been removed. A female lays up to one million eggs. The eggs settle to the bottom, stick to the sand, and are guarded and fanned by the male. The clearing of organic debris and silt from the nest site reduces competition for available oxygen between the eggs and decaying organic matter and other organisms on the bottom as well as helping to provide a constant flow of clean, oxygenated water to the embryos. The fry are guarded by the male for a few days until a school of independent small bass leaves the nest area. They then feed on small invertebrates, and when they are larger, they include fishes. Adults eat fishes, frogs, small ducklings, and almost any other animal of appropriate size. The largemouth bass matures in its second year and can reach a maximum of 13 years.

This is one of the most sought-after sport fishes in the mid-Atlantic region because it reaches record sizes in excess of 10 pounds in all states of the region, because of its presence in a variety of warmwater habitats, and because of its pugnacious behavior. It is sought primarily by baitcasting, with live minnows, worms, crickets, and other species, and by lures such as crankbaits, plastic worms, spinners, and topwater plugs.

White crappie

Pomoxis annularis Pl. 152

Description. 6.7 to 20.9 inches (170 to 530 mm). The white crappie has a small head and a large mouth and is intermediate in body shape between the elongate basses and the deep-bodied bluegill. It has six spines in the dorsal fin, while the similar-appearing black crappie has seven or eight, and the side has a series of dark bars, while that of the black crappie is dark mottled. The dorsal, anal, and caudal fins of the white crappie are weakly speckled with black and white, while in the black crappie they are much more heavily speckled with black and white.

Distribution and Abundance. The white crappie is native to the Mississippi River basin, but not, apparently, to the mid-Atlantic region. It has, however, been widely stocked and now occurs sporadically throughout the region except for the lower two-thirds of the Delmarva Peninsula. It is usually much less common than the black crappie.

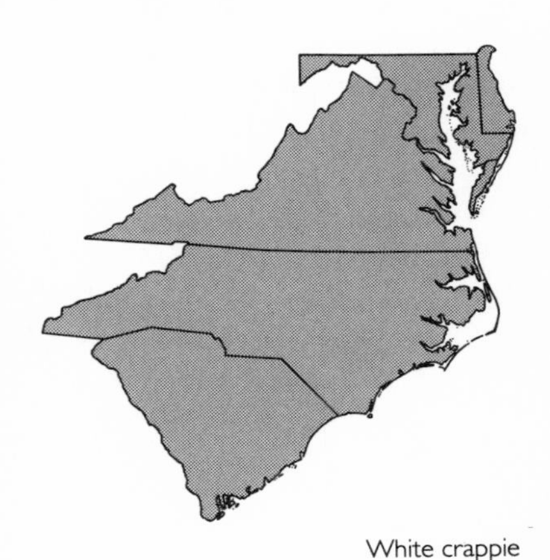

White crappie

Habitat. Both crappies favor slow waters, but the white crappie is adapted to warmer and more turbid waters than the black crappie. Both species are usually associated with aquatic vegetation and underwater shelter such as fallen tree tops.

Natural History. The white crappie builds a weakly defined nest. The female lays up to 213,000 eggs. Other aspects of reproduction appear to be similar to that of the black crappie (see the black crappie account below). Maturity is reached at an age of two to three years, and the maximum age reached is approximately nine years. Food is primarily small fishes, insects, freshwater shrimp, and amphipods.

Black crappie

Pomoxis nigromaculatus Pl. 153

Description. 5.1 to 19.3 inches (130 to 490 mm). The black crappie, often called speckled perch, has a small head, large mouth, and seven or eight spines in the dorsal fin. The body is silvery, with many small dark blotches on the side, and the dorsal, anal, and tail fins are heavily spotted with black and white.

Distribution and Abundance. The black crappie is native to most of the eastern half of the United States, and in the mid-Atlantic region to both the Carolinas and southeastern Virginia. It has been widely introduced and is now found throughout much of the region as well as in much of the rest of the United States. It is often common.

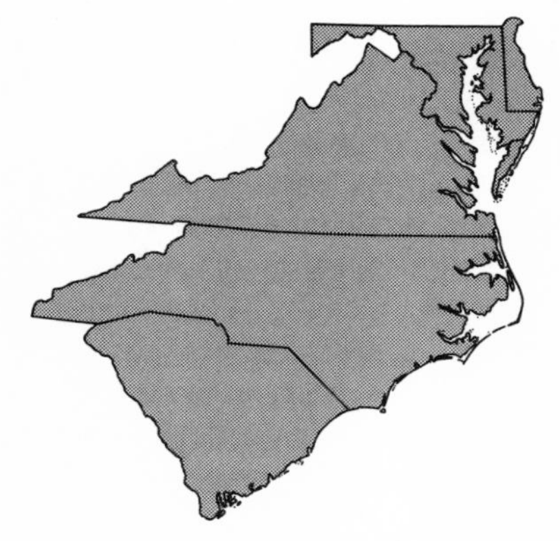

Black crappie

Habitat. An inhabitant of slow black water and nonstained streams of the coastal plain, it prefers clear and cooler areas, and it does not grow well where heavy siltation causes turbidity. It is usually found around underwater logs and broken tree tops. It has been widely introduced into ponds and lakes.

Natural History. The black crappie builds a nest typical of sunfishes, but often not as well-defined, usually in shallow, calm water near vegetation. The nests are often grouped, and each is guarded by the male until the eggs hatch and the young leave. A female spawns several times in one season, which extends from spring to summer, and can lay up to 188,000 eggs. The food is aquatic insects and small fishes when the black crappie is young and primarily fishes when it is adult. It is an important game fish and provides fishing opportunity in late winter and early spring before other species feed. It is an excellent food fish. Fish caught usually weigh less than one pound, but they will occasionally reach four pounds.

Pygmy Sunfishes *Family Elassomatidae*

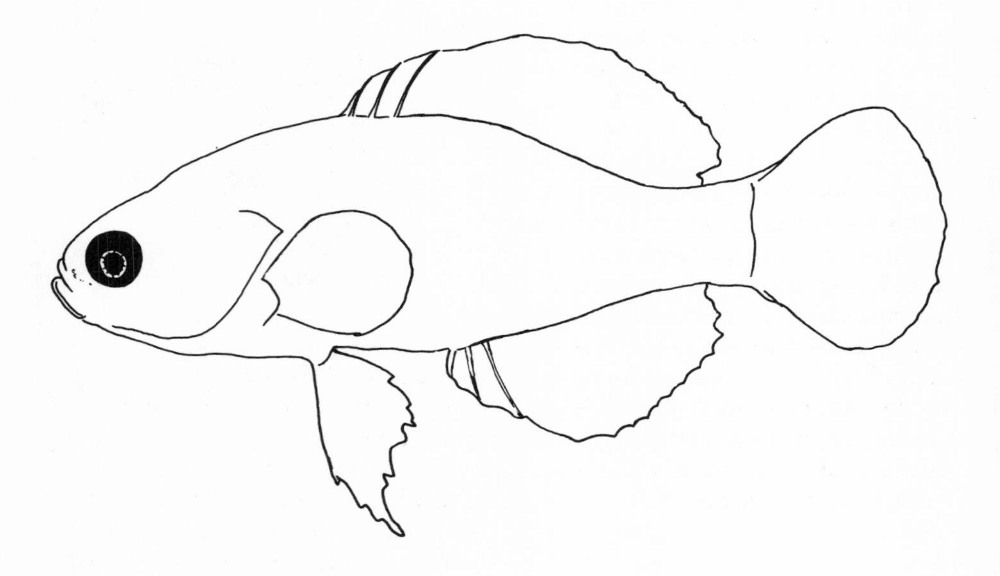

Pygmy sunfishes are small and secretive and are found only in the southeastern United States. There are six known species, all in one genus. Long considered to be members of the family of sunfishes (the Centrarchidae), the pygmy sunfishes are so distinct in their anatomical, behavioral, and genetic characteristics that many ichthyologists now place them in their own family, the Elassomatidae. *Elassoma* is derived from a Greek word and means "small."

Carolina pygmy sunfish
Elassoma boehlkei Pl. 154
Everglades pygmy sunfish
Elassoma evergladei Pl. 155
Bluebarred pygmy sunfish
Elassoma okatie Pl. 156
Banded pygmy sunfish
Elassoma zonatum Pl. 157

Description. All members of this family are less than two inches long, lack a lateral line, and have a rounded caudal fin. Four species are found in the mid-Atlantic region.

Carolina pygmy sunfish: 0.8 to 1.3 inches (20 to 32 mm). A series of alternate black and blue bars of equal width appears on the side of the male. The bars of the female, however, are alternately dark brown and light brown.

Everglades pygmy sunfish: 0.9 to 1.3 inches (23 to 32 mm). The body of both sexes can be plain, mottled, or streaked. Scattered blue spots and flecks occur only on the body and head of the male.

Bluebarred pygmy sunfish: 0.9 to 1.4 inches (24 to 35 mm). This species appears similar to the Carolina pygmy sunfish, but the black bars are approximately three times as wide as the bright blue bars.

Banded pygmy sunfish: 1.0 to 1.8 inches (25 to 45 mm). This is the largest pygmy sunfish. Identifiers are a dark blotch (although sometimes there are two or three) on the shoulder and both a dark stripe behind and a dark bar below the eye. Both sexes have distinct bars of brown or black, which in the male alternate with bars of green or gold and in the female with bars of gray.

Distribution and Abundance. In the mid-Atlantic region, pygmy sunfishes are found only in the coastal plain of the Carolinas. The banded and the Everglades pygmy sunfishes are widespread in the region, and both also occur in other portions of the United States. The other two species, which were only recently recognized as new, are restricted to parts of the region. Pygmy sunfishes are often abundant in their preferred habitat. Because of their limited distributions, the Carolina pygmy sunfish is considered threatened in North Carolina and the bluebarred pygmy sunfish is listed as of special concern in South Carolina.

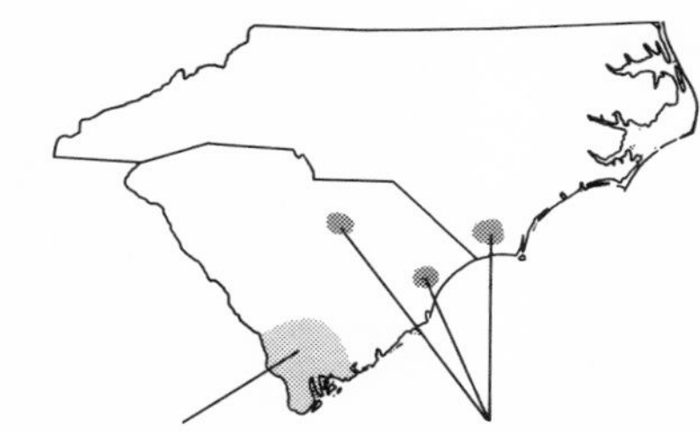

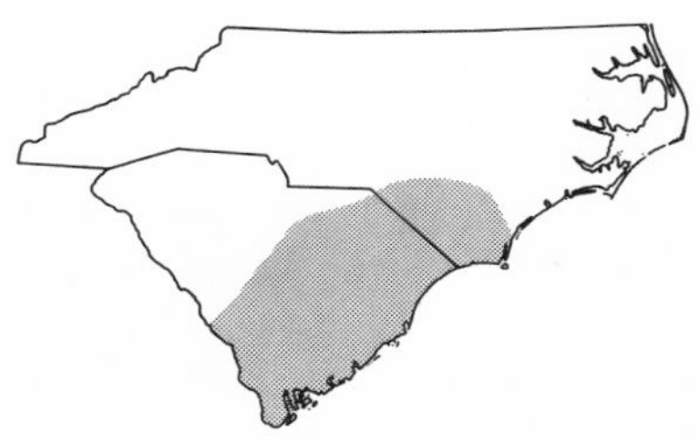
Everglades pygmy sunfish

Habitat. Pygmy sunfishes inhabit calm and usually tannin-stained, acidic waters of roadside ditches, sloughs, small ponds, and portions of creeks. They occur in heavy vegetation in shallows.

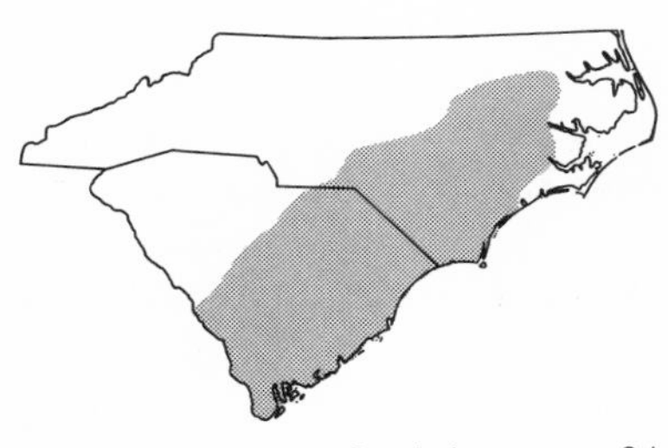
Banded pygmy sunfish

Natural History. All pygmy sunfishes have a similar life history. Although often common, they are not conspicuous because they are solitary, secretive, and often motionless in or beneath aquatic vegetation. They do not school and they are not active at the surface, unlike the killifishes and poeciliids that usually occur with them. The males become colorful during the breeding season in early to mid-spring. Courtship and breeding occur among vegetation; they are not known to construct a nest. Up to 76 eggs per female are deposited on fine-leaved plants; after they hatch, the larvae grow quickly and mature within a year. Few live longer than 18 months. The diet is primarily of microcrustaceans, such as copepods and cladocerans, and aquatic insects, primarily fly larvae. If provided with clean and slightly acidic water and live food (they rarely eat prepared food), their culture in aquariums is easy and they can readily be induced to spawn.

Perches *Family Percidae*

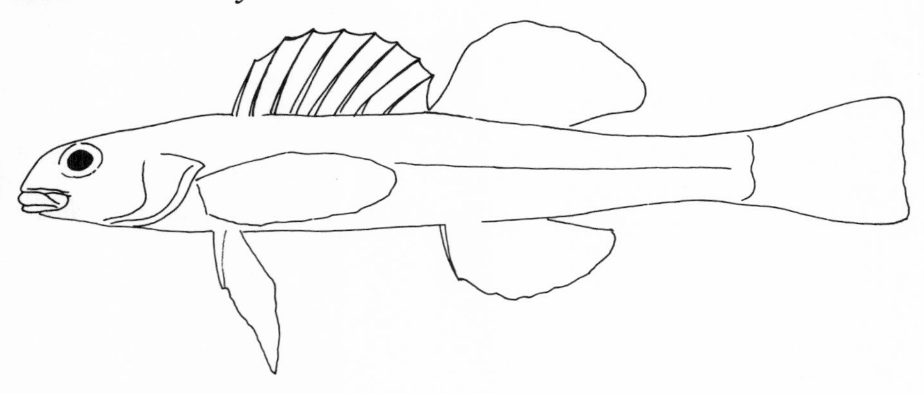

The perches are characterized by two dorsal fins, which are separate or weakly joined, and by one or two spines in the anal fin, of which the second, if present, is slightly longer than the first. The pelvic fins have one spine and five soft rays and are located in the breast area. Perches are freshwater fishes occurring in most of eastern and central North America, almost all of Europe, and most of northern Asia. The family contains a total of approximately 160 species. About 150 of these are darters, which are small, usually colorful bottom dwellers and are restricted to North America. The greatest species diversity occurs in the southern Appalachian Mountains. The largest species is the walleye, which reaches a length of approximately 35 inches, and the smallest is a darter. This family also includes the widely distributed and well-known yellow perch.

Western sand darter

Ammocrypta clara

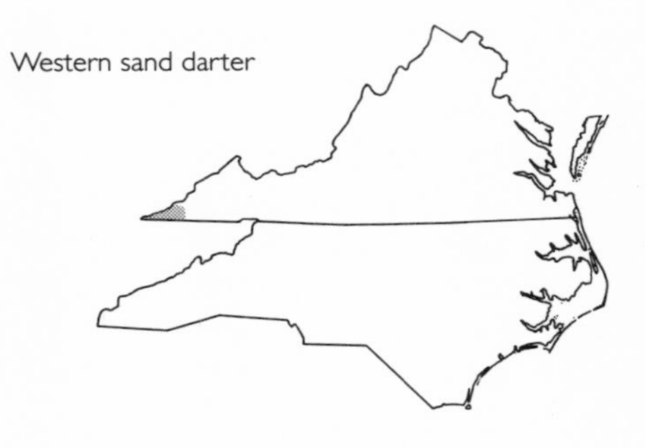

Description. 1.6 to 2.8 inches (41 to 71 mm). The body is extremely slender, translucent, and light yellow, the eye is located near the top of the head, and there is usually a row of dusky spots along the back and another row on the side.

Distribution and Abundance. The western sand darter occurs within the mid-Atlantic region only in extreme western Virginia, in which state it is rare and is listed as threatened. Several small to large disjunct populations of it occur in the central United States from Wisconsin to Texas.

Habitat. This darter occurs in medium-sized and large rivers with a sand or sand/small gravel substrate. It prefers a low to moderate gradient and warm water. Like several close relatives in the genus *Ammocrypta,* it often lies buried in the substrate with only its eyes and mouth exposed, or conceals itself totally, for which behaviors the characteristics given in the description above are adaptations. This unusual burrowing may be an adaptation to escape from predators, to conserve energy (by not having to swim against a current and by providing a lower body temperature under sand), and/or a type of lie-in-wait predation in which prey is caught in a dash from under sand.

Natural History. The western sand darter eats primarily larvae of aquatic insects. It breeds in May and June in the mid-Atlantic region and several weeks later farther north. The adults then presumably move into shallow riffles. The ripe egg is small, and each female produces from 60 to 300. Reproductive maturity is probably reached in one year, and the sand darter reaches a maximum age of three years. This species apparently occurs at a low population density even under normal conditions. It cannot tolerate siltation or water level fluctuation, and consequently it has been extirpated or greatly reduced in numbers over large parts of its range.

Greenside darter
Etheostoma blennioides Pl. 158
Turquoise darter
Etheostoma inscriptum Pls. 159–60
Kanawha darter
Etheostoma kanawhae Pl. 161
Maryland darter
Etheostoma sellare
Swannanoa darter
Etheostoma swannanoa Pl. 163
Seagreen darter
Etheostoma thalassinum Pl. 164
Banded darter
Etheostoma zonale Pl. 165

Description. The subgenus *Etheostoma* contains 14 species, 9 of which occur in the mid-Atlantic region. Morphological variation in this group is considerable, but members generally have a complete and straight lateral line, well-separated pelvic fins, a broad union of the gill membranes, a blunt snout, large pectoral fins that are longer than the short head, and dark saddlelike blotches on the back.

Greenside darter: 3.1 to 6.5 inches (78 to 166 mm). This is the largest darter in the genus *Etheostoma.* The skin at the rear of the upper lip is fused to the blunt snout, and it has green W- or U-shaped markings on the side. The overall color is a yellow-green. The male can develop green bars on the side and fins and a red base on the dorsal fin.

Turquoise darter: 2.6 to 3.1 inches (66 to 78 mm). This darter is one of a distinctive subgroup with the Swannanoa and seagreen darters. Adults of both sexes are characterized by six dark saddles on the back and six blotches on the side, but the side of the male is brown with several horizontal rows of small red spots, while the side of the female is yellow-brown with brown saddles and blotches. The breeding male has a pinkish body with dark green bars on the side and blue on the lower part of the head, blue on the lower portions of the pelvic and anal fins and on the outer edges of the caudal fin, blue-green on the base of the dorsal fins and red on the edges, and red on the caudal fin.

Kanawha darter: 2.6 to 3.5 inches (66 to 90 mm). This is one of the region's most colorful darters, with five or six large, dark saddles on the back and broad green bars on the side, alternating with thinner orange bars; the breeding male has an orange belly and gill area. The **candy darter**, *E. osburni* (Pl. 162), is nearly identical but has 58 to 70 lateral scales versus 48 to 58 in the Kanawha darter, and its red or orange and blue-green bars are more brilliant: its appearance is "candy-striped." The **variegate darter**, *E. variatum*, is similar to both, but it has only four large dark brown saddles.

Maryland darter: 2.1 to 3.3 inches (54 to 84 mm). This fish is characterized by a wide and flat head, four large, dark saddles, and narrowly joined gill membranes. It is brown to yellow olive and less colorful than the other members of this subgenus.

Swannanoa darter: 3.1 to 3.5 inches (78 to 90 mm). This darter has six black saddles on the back and has eight or nine oval to round black blotches, along with rows of small red spots, on the side. The edge of the first dorsal fin is red, and the anal and pelvic fins in the breeding male become bluish-green. The female is marked similarly, but the colors are subdued.

Seagreen darter: 2.6 to 3.1 inches (66 to 78 mm). Similar to the Swannanoa darter, this species has seven dark brown saddles and only seven small dark brown blotches on the side. The first dorsal fin has a red edge and a dusky black base, and the second dorsal fin is reddish-orange. The pelvic and anal fins are blue.

Banded darter: 2.1 to 3.1 inches (54 to 78 mm). The male of this colorful darter has nine or more large dark green bars on the side, and these extend down to the belly; these bars on the female and juvenile are yellowish-tan. The first dorsal fin is edged with green lying above a red band.

Distribution and Abundance. The greenside darter in the mid-Atlantic region occurs in the mountains from North Carolina to Maryland; it also occurs throughout most of the eastern United States from the Ozark Mountains in Arkansas northeast to New York. The banded darter is found in the mountains of North Carolina and Virginia, and there is an introduced population in northeastern Maryland; it also occurs in most of the eastern United States. The other seven species have very restricted distributions. The Swannanoa darter is confined to the mountains of North Carolina, Virginia, and Tennessee in tributaries of the Tennessee River. The seagreen darter is found in lower elevations in tributaries of the Santee River drainage in the mountains and in the upper piedmont of North Carolina and South Carolina. The turquoise darter is found primarily in similar areas in southwestern North Carolina, South Carolina, and Georgia (Edisto, Savannah, and Altamaha drainages). Because of its limited distribution in North Carolina, it is listed as of special concern by that state. The Kanawha darter is endemic to the upper New River drainage in North Carolina and Virginia, while the candy darter is restricted to the New River drainage in Virginia and West Virginia. The variegate darter is found in the Ohio River basin and, within the mid-Atlantic region, occurs only in southwestern Virginia. The Maryland darter is currently known from only one riffle in one creek in eastern Maryland, although in recent years it was known from two others. It is one of the rarest fishes in North America and is listed as endangered by the federal government. With the exception of the Maryland darter, all members of this subgenus are common in appropriate habitat.

Habitat. The preferred habitat of all nine species is similar: all favor rock, rubble, or gravel riffles with a moderate to swift current, and the group occurs in creeks and small to medium-sized rivers.

Natural History. Little is known about the biology of these species other than

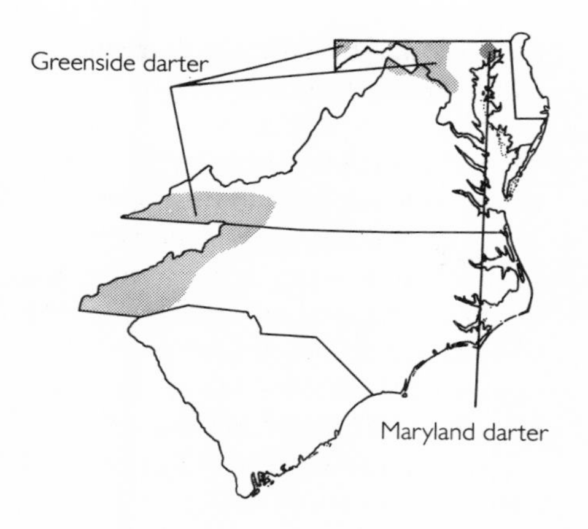

Banded darter

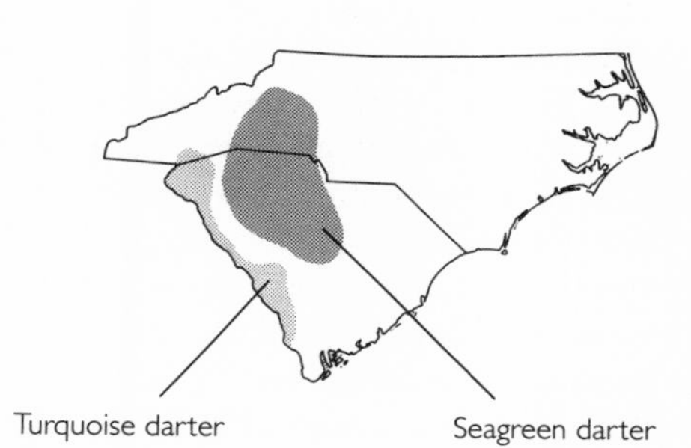

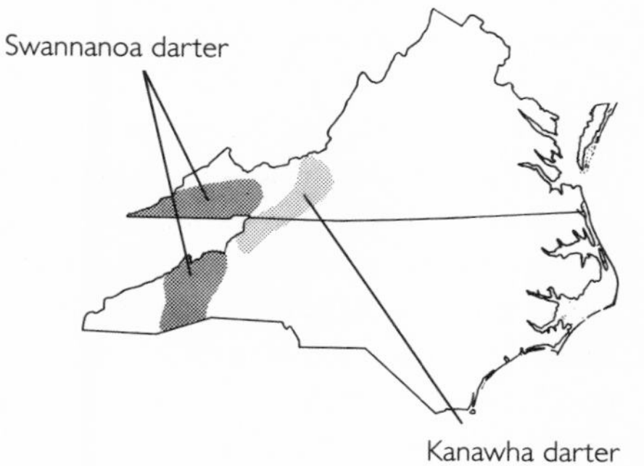

the greenside darter. The greenside and banded darters spawn in late spring. Others in the group probably spawn from mid-spring to early summer. The male greenside darter is larger than the female, and he establishes and defends a territory. The eggs are attached to strands of filamentous algae and aquatic mosses; after they hatch, growth is rapid and maturity is reached by the end of the first year of life. The maximum age is four years. Dominant foods are larvae of aquatic insects. The Maryland darter prefers snails and caddisfly larvae.

Longfin darter
Etheostoma longimanum
Johnny darter
Etheostoma nigrum Pl. 166
Tessellated darter
Etheostoma olmstedi Pl. 167
Waccamaw darter
Etheostoma perlongum Pl. 168
Riverweed darter
Etheostoma podostemone Pl. 169

Description. Five species are included in the subgenus *Boleosoma,* all of which occur in the mid-Atlantic area. The johnny, tessellated, and Waccamaw darters are similar, while the longfin and riverweed darters form another group. All five species are characterized by a relatively blunt snout, a series of X's or W's along the midside of the body, a straight and complete lateral line, a wide interspace between the pelvic fins, and a groove that separates the upper lip from the snout. All five species are also characterized by a lack of breeding tubercles. Further, the female deposits eggs on the underside of stones or logs with a flat and bifurcate (forked) genital papilla.

Longfin darter: 1.6 to 3.5 inches (42 to 89 mm). This is a robust species with relatively long pectoral fins in both sexes and a higher and longer second dorsal fin than first dorsal fin in the male. There is a row of 9 to 14 dark square blotches or W marks on the side. The body of the breeding male has an orange cast, and the dorsal fins are mostly orange, and black along the base.

Johnny darter: 0.9 to 2.8 inches (24 to 72 mm). The johnny darter is light brown or straw-colored above with six dark brown saddles, a series of black X's or W's on the side, and dark wavy flecks on the upper side. The breeding male has a black head and anal and pelvic fins and white knobs on the tips of the rays of the pelvic fin.

Tessellated darter: 2.1 to 4.3 inches (53 to 110 mm). This species is very similar to the johnny darter, but the male has a higher and longer second dorsal fin than the johnny darter.

Waccamaw darter: 2.1 to 3.5 inches (53 to 90 mm). Quite similar to the tessellated darter, but the Waccamaw darter is more slender and the breeding male has reddish dorsal fins.

Riverweed darter: 1.6 to 3.0 inches (42 to 76 mm). Similar to the longfin darter, it is distinguished by rows of dark spots on the side and the caudal peduncle and by a distinct dark bar, not a spot, below the eye.

Distribution and Abundance. The longfin darter is restricted to the upper James River drainage in Virginia and West Virginia, where it is common. The johnny darter is widely distributed from Canada to Alabama and Mississippi, west to Colorado, and east to Maryland, Virginia, and North Carolina, where it is common to abundant. The tessellated darter is common to abundant in the piedmont and coastal plain throughout the mid-Atlantic region; it also occurs in Atlantic slope drainages from the St. Lawrence River in Quebec and Ontario south to the St. Johns River in Florida. The Waccamaw darter is found only in Lake Waccamaw and the headwaters of the Waccamaw River in southeastern North Carolina, where it is moderately common. The riverweed darter is restricted to the upper Roanoke River drainage of Virginia and North Carolina, where it is common. In North Carolina the Waccamaw darter is considered threatened and the riverweed darter is listed as of special concern.

Habitat. The longfin and riverweed darters prefer rock riffles of creeks and small rivers. The johnny and tessellated darters inhabit sand- and mud-bottomed pools of headwaters, creeks, small to medium-sized rivers, and shores of lakes, and the johnny darter sometimes occurs over a rock substrate. The Waccamaw darter occurs along sandy shorelines in waters less than six feet deep.

Natural History. All five species are spring and early summer spawners in the mid-Atlantic region. The male fans away sand with the caudal fin to prepare a nest under a flat rock, stick, log, or other cover. He there courts a female as he shows his fully extended fins in a lateral display, at times also showing the darker and sometimes more colorful breeding colors of his head and dorsal fins, and drives off any male that approaches. A female ready to spawn enters the nest, after which she and the male assume an upside down position and deposit eggs and sperm on the underside of the nest cover. A female typically deposits a few dozen to several hundred eggs during a single spawning session and then leaves. The male guards the eggs and courts subsequent females. The underside of the nest cover may become coated with a single layer of several thousand eggs. The eggs hatch in about a week, and no care is given to the young. Longevity ranges from one year for the Waccamaw darter to over four years for the johnny darter. These species reach breeding age in their first or second year, by which time most of their adult size has been attained. The male grows larger and faster than the female. Food is midge larvae, mayflies, and small crustaceans. These darters are easily maintained in aquariums and readily take both live and commercial food.

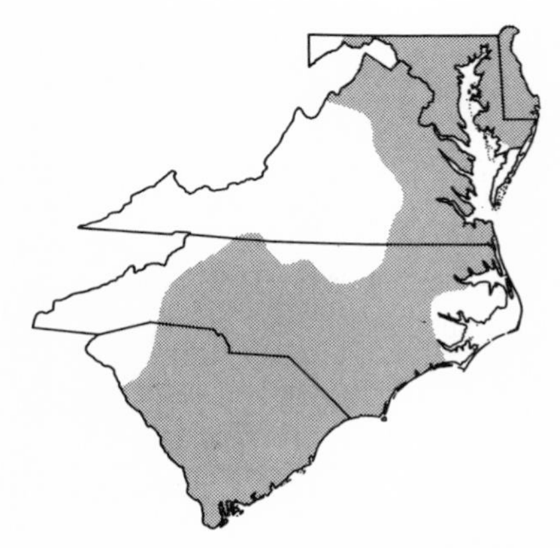
Tessellated darter

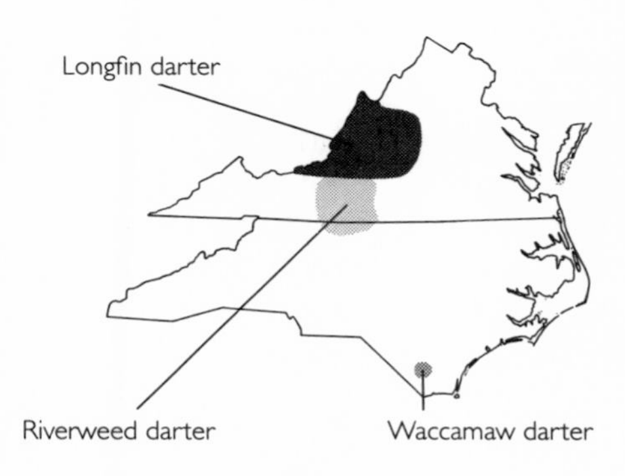

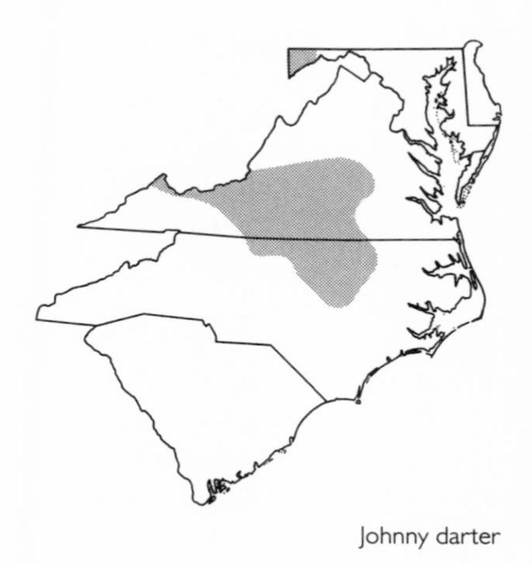
Johnny darter

Glassy darter

Etheostoma vitreum Pl. 170

Description. 1.8 to 2.6 inches (47 to 66 mm). The glassy darter is the only member of the subgenus *Ioa.* It is most closely related to the members of the subgenus *Boleosoma* and superficially resembles the sand darters, genus *Ammocrypta.* The body is slender and translucent, and the snout is pointed. There are seven or eight tan or brown saddles, the back and side are speckled and spotted in brown and black, and there are dark dashes or V's along the midside. The breeding female is dusky and the male sooty. The anus is surrounded by small villi, and both sexes have breeding tubercles on the pectoral and pelvic fins.

Distribution and Abundance. The glassy darter is restricted to the mid-Atlantic region, where it occurs in the upper coastal plain and piedmont from the Bush River in Maryland to the Neuse River in North Carolina. It was recently collected in the central portion of the Delmarva Peninsula. This species is often common, except in Maryland, where it is considered endangered.

Habitat. This species occurs in sand and gravel runs of creeks and small to medium-sized rivers with a moderate to fast current.

Natural History. The glassy darter usually remains buried in sand and emerges to forage at dawn or dusk or when it is disturbed. Its food is probably insect larvae and tiny crustaceans. This is the only darter for which communal spawning has been reported. It spawns in March and early April at locations with a steady current. Breeding areas are occupied by many males that nudge or butt one another aggressively until females arrive one or two at a time. Eggs are deposited in a series of passes over a communal nest located upstream of rocks, beneath logs, or on concrete spillways. A female, closely accompanied by one or two males, deposits about 100 eggs in these passes. She spawns only once per year. An active nest attracts additional individuals and may accumulate 30,000 to 50,000 eggs.

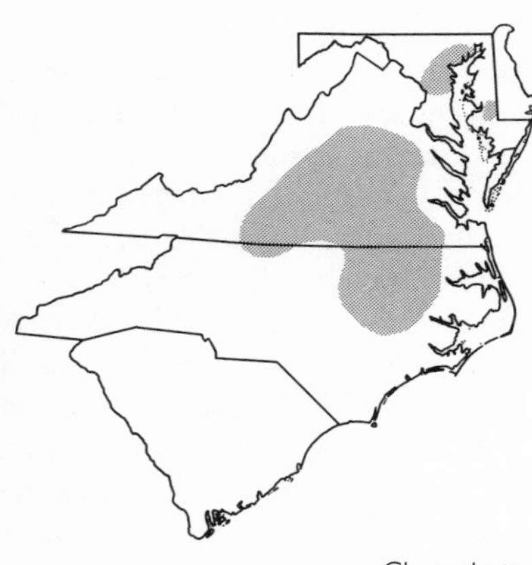

Glassy darter

Snubnose darter

Etheostoma simoterum Pl. 171

Description. 1.9 to 3.0 inches (48 to 77 mm). The 15 members of the subgenus *Ulocentra* are referred to as the snubnose darters. They are characterized by a blunt snout, broadly joined branchiostegal membranes, and no, or only a narrow, frenum. The snubnose darter, the one species of the group found in the mid-Atlantic region, is beautiful and marked with many small red spots in short rows on the upper side. The first dorsal fin has a red edge, wavy black lines, and a red spot at the front; there are eight or nine dark saddles on the back and eight or nine black blotches on the side. The breeding male has a bright orange belly, a bluish snout, and bluish anal and pelvic fins.

Distribution and Abundance. This fish occurs in the Tennessee River drainage from northern Alabama to southwestern Virginia, and it is one of the most abundant darters in its range, occurring at a density of up to five individuals per square meter. However, it has apparently been extirpated from North Carolina, where it is now listed as of special concern, although a population might still occur near the Tennessee border.

Habitat. It prefers small and clear creeks with a moderate current over a gravel substrate, but it also occurs in larger creeks and moderate-sized rivers, sometimes over rubble and bedrock. It is usually found in backwaters and calm shallows

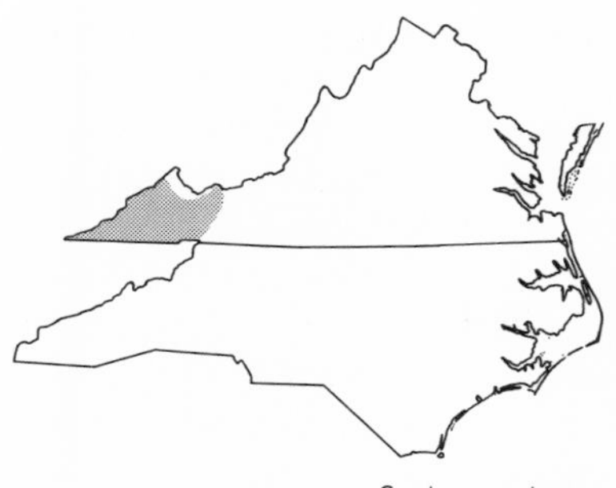
Snubnose darter

Natural History. The snubnose darter spawns in April and May. The male is not territorial but is combative when in close proximity to another male. The female is courted by the larger male, which then becomes highly colorful and laterally displays his median fins. When ready to spawn, the female leads the male to a particular site, usually a large stone; he then mounts her, and one or two eggs are deposited on the side or top of the stone. The pair then moves off to another site and repeats the act. Up to 240 eggs are laid by a female. However, there is much switching of partners. Longevity is up to 18 months. Food is primarily midges, but a wide variety of crustaceans and aquatic insects is consumed.

Blueside darter

Etheostoma jessiae Pl. 172

Description. 1.9 to 3.0 inches (49 to 77 mm). One of two species in the subgenus *Doration*, the male blueside darter is spectacularly colored and can be recognized by 7 to 11 blue bars on the side, blue on the face, and orange spots on the tail and pectoral fin. There is an orange band across the middle of the first dorsal fin and a light blue stripe on the edge. There are 7 to 11 brown squares or W's on the side below the usually incomplete lateral line on the female and the juvenile.

Distribution and Abundance. This darter occurs only in the middle and upper Tennessee River drainage in North Carolina (Mills River system), Virginia (Clinch, Powell, and North Fork Holston rivers), Alabama, Georgia, and Tennessee. It is generally common. It may, however, have been extirpated from North Carolina and Virginia, and it is listed as a species of special concern in the former state.

Habitat. This species prefers moderate-sized creeks with a sand, gravel, or rock substrate, where it occupies both riffles and pools over a sand and detritus bottom. It prefers a slow to moderate current.

Natural History. Little is known about this darter. It spawns in March and April, usually in shallow riffles. The male mounts the female, and the fertilized eggs are buried in gravel, several at a time. The male apparently defends a shifting territory. This species is highly intolerant of habitat alteration, such as eutrophication from agricultural runoff and siltation.

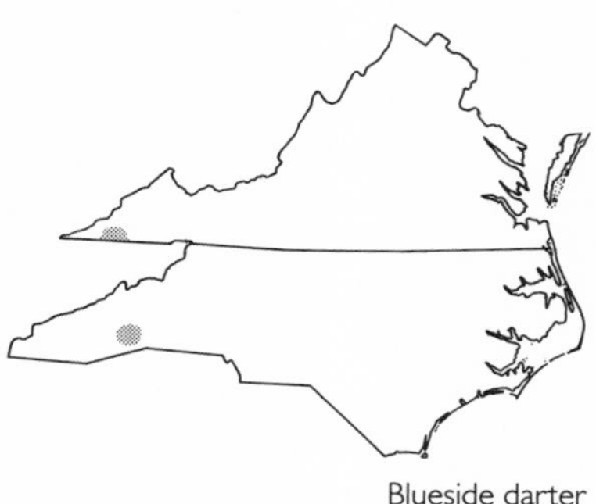
Blueside darter

Sharphead darter

Etheostoma acuticeps Pls. 173–74

Greenfin darter

Etheostoma chlorobranchium Pls. 175–76

Redline darter

Etheostoma rufilineatum Pl. 177

Wounded darter

Etheostoma vulneratum Pl. 179

Description. Six members of another grouping of darters, the subgenus *Nothonotus,* occur in rivers in the mid-Atlantic area. All have a distinctive slab-sided body with thin alternating dark and light stripes on the side and a pointed snout. The membranes of the first dorsal fin are dark.

Sharphead darter: 1.6 to 3.3 inches (42 to 84 mm). As implied by the common name, the snout is strongly pointed. Dusky bars are present on the side. The body color of the female is yellowish-brown, while in the breeding male it is olive to blue. The **Tippecanoe darter,** *E. tippecanoe* (Pl. 178), differs from all other members of this subgenus by the presence of a large black bar that encircles the caudal peduncle. The breeding male has a golden orange body and fins. It is also a tiny fish, less than 1¾ inches long.

Greenfin darter: 2.1 to 3.9 inches (54 to 100 mm). The large breeding male develops a deep green body and deep green fins, except for the pectoral fin, which becomes pink. The female, juvenile, and nonbreeding male are brown with small black and reddish-brown spots on the side. The dorsal, caudal, and anal fins are edged with black. The snout is rounded, in contrast to the other species in this subgenus. The **bluebreast darter,** *E. camurum,* is similar to the greenfin darter, but the fins are amber or slightly orange and the breast blue or gray.

Redline darter: 2.0 to 3.3 inches (52 to 84 mm). This is one of the mid-Atlantic region's more colorful darters and is identified by black dashes on the operculum. The male has red spots on the side and a red-orange band below the black edge of the fins. The body of the female is patterned similarly to that of the male but is less colorful, and the first dorsal fin and the anal fin of the female are yellowish with many dark spots. The base of the caudal fin in both sexes is cream-colored.

Wounded darter: 1.9 to 3.1 inches (48 to 80 mm). The body of this darter is strongly compressed, and its overall color is gray, with a black edge on the second dorsal, caudal, and anal fins. Red spots are present in the front and rear of the first dorsal fin, and the second dorsal and caudal fins are also partly red. The red spots on the side of the male are surrounded by a black circle.

Distribution and Abundance. Four of the six species are restricted to the Tennessee River drainage in the mid-Atlantic region, and except for the redline darter all are limited in distribution. The sharphead darter is the most localized and is known only from southwestern Virginia, extreme western North Carolina, and eastern Tennessee, from the upper Nolichucky River and two of its tributaries and from the South Fork Holston River. It is considered endangered in Virginia and threatened in North Carolina. The greenfin darter and the wounded darter often occur together in tributaries of the upper Tennessee River from Virginia to Georgia; the former is listed as threatened in Virginia, and the latter is listed as of special concern in North Carolina. The redline darter has the widest distribution; it is found in the Cumberland and Tennessee rivers from Virginia southwest to Mississippi, and in the mid-Atlantic area occurs in Virginia and North Carolina. The bluebreast darter and the Tippecanoe darter occur only in the Ohio River basin, where the latter is extremely localized; in the mid-Atlantic region both species occur only in southwestern Virginia. The Tippecanoe darter is considered threatened in Virginia. The sharphead darter is rare in Virginia, while the other five species are common in their preferred habitat.

Habitat. The bluebreast darter, sharphead darter, and wounded darter are restricted to fast and rocky riffles in small to medium-sized rivers. The Tippecanoe darter is restricted to "pea" gravel in swift water. The other two species also occur in these habitats, as well as in creeks.

Natural History. These species spawn in summer. Sexual dimorphism then is particularly pronounced, and the male of each species is much more colorful than the female. The female buries herself in gravel and then deposits several eggs; from 100 to over 400 eggs are deposited annually. Growth of hatchlings is rapid. Maturity is reached by the end of the first year, and the maximum age attained is three years. The dominant food is larvae of aquatic insects, particularly flies.

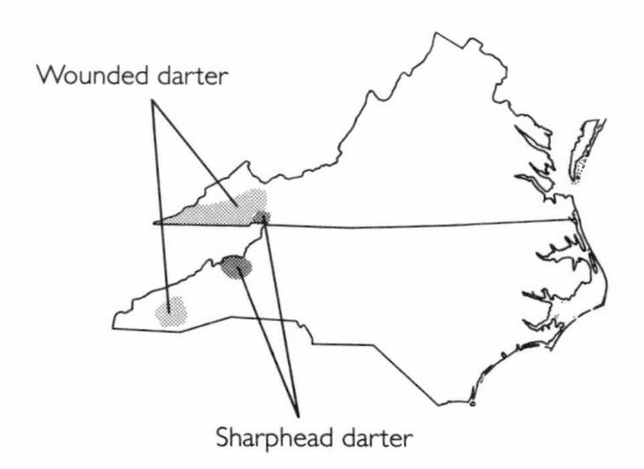

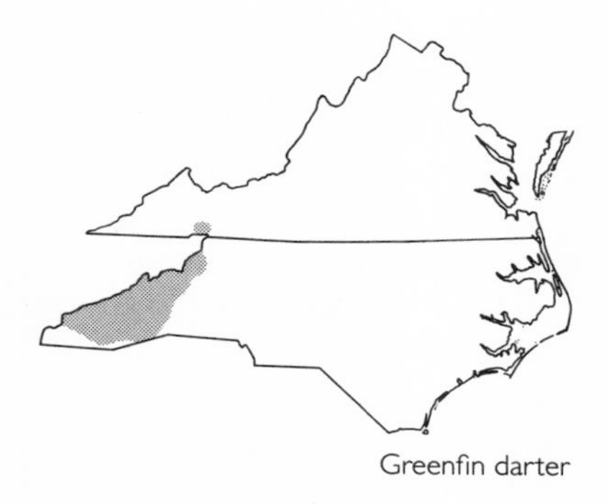

Greenfin darter

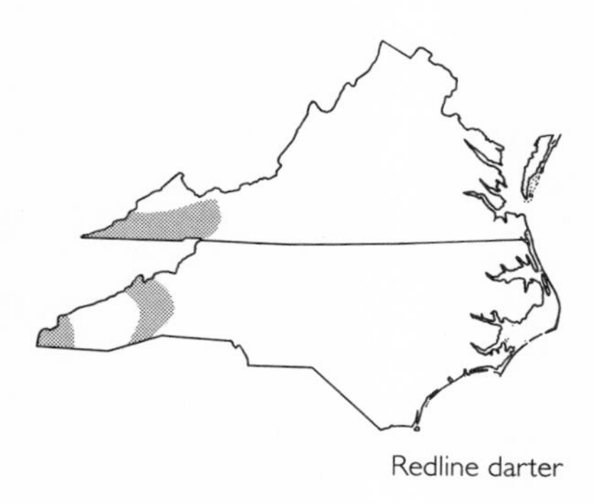

Redline darter

Savannah darter
Etheostoma fricksium Pls. 180–81
Pinewoods darter
Etheostoma mariae Pl. 182

Description. These two species are placed in the subgenus *Belophlox* and can be separated from all other darters by the presence on the side of a broad dark stripe and a contrasting light-colored lateral line. Further, a light brown back contrasts with a darker side in both species. The margin or submargin of the first dorsal fin is red.

Savannah darter: 1.4 to 2.9 inches (36 to 74 mm). It is characterized by green bars that alternate with reddish-orange bars on the lower side and by orange on the belly. The interior of the first dorsal fin is bluish, and the second dorsal fin and caudal fin are heavily dark speckled.

Pinewoods darter: 1.9 to 3.0 inches (48 to 76 mm). It lacks the red and green bars of the Savannah darter. There is a black spot at the front of the first dorsal fin and many black spots on the lower part of the head and breast. The lateral line is yellowish and conspicuous.

Distribution and Abundance. Both of these darters have a restricted distribution. The Savannah darter is found only in the upper coastal plain in south central South Carolina and adjacent eastern Georgia. The pinewoods darter is restricted to the sandhills area just below the fall line in the Little Pee Dee River, which includes the Lumber River in south central North Carolina. There is an old record from adjacent South Carolina, and, based on it, the pinewoods darter is listed as of special concern in that state. Both fishes are uncommon to moderately common in appropriate habitat.

Habitat. Both species are characteristic of small to medium-sized creeks, usually in a pronounced current over sand and gravel. The pinewoods darter is typically associated with submerged aquatic vegetation. The Savannah darter is also found in such vegetation and in exposed submerged tree roots and debris of undercut stream banks.

Natural History. Both species spawn in the spring, with a peak in April and May. The Savannah darter attaches its eggs to sand and gravel, while the pinewoods darter may bury them in gravel. The female of the pinewoods darter contains an average of 60 mature eggs, which are probably deposited in several clutches rather than in one. Growth is rapid in both species. Maturity is reached at the end of the first year of life, and the maximum age reached is three years.

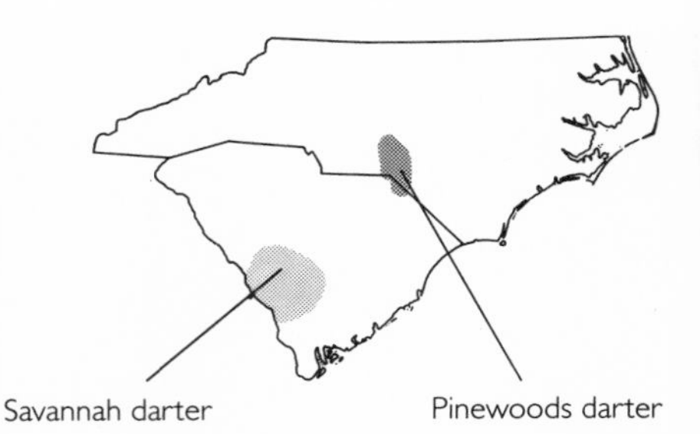

Common foods are aquatic larvae of insects such as midges, flies, mayflies, and stoneflies.

Christmas darter

Etheostoma hopkinsi Pl. 183

Description. 1.4 to 2.6 inches (35 to 66 mm). The colorful Christmas darter is characterized by 10 to 12 dark green bars on the side, each separated by an orange-red bar in the mature male and a yellow bar in the female. The dorsum is greenish, with eight darker saddles, and the venter is light green. The first dorsal fin is bordered by a thin green edge, and there is a red band in the middle. The **rainbow darter**, *E. caeruleum*, has blue and red bars on the side, and its dorsal, caudal, and anal fins are red with a blue edge. These two are the only members of the subgenus *Oligocephalus*, which contains a total of 17 species, that occur in the mid-Atlantic region.

Distribution and Abundance. The Christmas darter is found in west central South Carolina and in eastern and central Georgia. In the mid-Atlantic region the rainbow darter occurs only in western Virginia. It is widespread to the west. Both are often locally abundant.

Habitat. Rock or gravel riffles of creeks and small to medium-sized rivers, in a

Christmas darter

medium to fast current, are the Christmas darter's habitat. It is occasional in slow areas and in submerged vegetation.

Natural History. Little-known, the Christmas darter lives behind and under stones and feeds on small aquatic invertebrates, primarily insects, especially their larvae. It probably reproduces in March and April. Sexual maturity is attained at an age of one year, and it probably reaches a maximum age of two to three years.

Fantail darter

Etheostoma flabellare Pl. 184

Description. 1.3 to 3.3 inches (33 to 84 mm). The fantail darter is a member of the subgenus *Catonotus*, which contains at least 12 species, all characterized by a broad and flat (not bifurcate) genital papilla in the female and by their behavior of attaching eggs to the underside of stones. The fantail darter is identified by six to eight distinct black bars on the tail, by more numerous but less distinct black bands on the second dorsal fin, and by black bars on the body. There are gold-colored knobs on the tips of the dorsal fin spines, and these are larger on the male than on the female. The breeding male is brownish, with the head and fins black. A closely related and undescribed species, the **duskytail darter**, has a greater number (10 to 15) of more narrow and less distinct bars on the body, and weaker bars or no bars on the tail.

Distribution and Abundance. In the mid-Atlantic region the fantail darter is found in extreme northern South Carolina, the western two-thirds of North Carolina, and the western half of Maryland and Virginia. It occurs widely in other portions of the eastern United States. It is often abundant. The duskytail darter is restricted to extreme western Virginia and adjacent Tennessee, where it is generally rare; it is considered endangered in Virginia.

Habitat. The fantail darter is usually found in rock and gravel riffles of medium- to fast-flowing creeks and smaller rivers. Juveniles can be found in all habitats.

Natural History. The fantail darter male is larger than the female. It probably spawns from March to June in the mid-Atlantic region. When the water temperature reaches 45° to 57°, the male establishes a stationary territory. He then clears debris from the underside of a flat rock, excavates underlying gravel, and

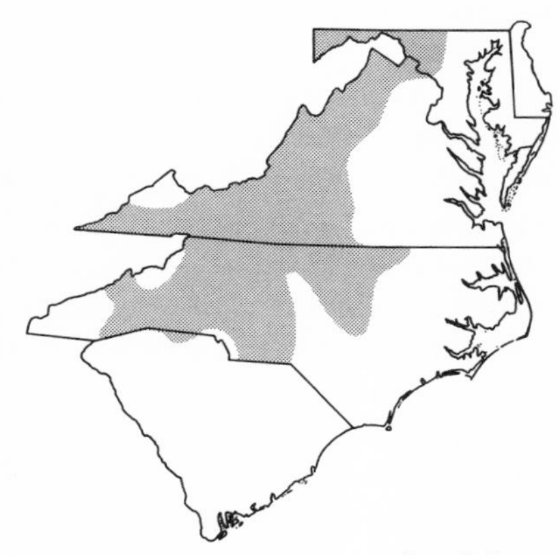
Fantail darter

defends the site against any fish of his size. He becomes almost black when he chases off an interloper. Females are accepted. The duration from courtship to spawning is about one minute. The eggs are fastened to a rock undersurface, and the fertilizing male accompanies the female in an inverted position. Both sexes spawn with several partners in one season, and a nest may contain several hundred eggs. Egg deposition and spawning are repeated every one to three minutes until the female deposits about 35 eggs. She deposits a total of over 200 eggs in one reproductive season. The male defends the eggs against predators such as crayfishes and sculpins, cleans them by removing debris, and fans them with his fins to aerate them. They hatch in 14 to 35 days, depending on the water temperature. The hatchlings are a quarter inch long, grow rapidly, and can probably reproduce the next year. Food is small insects, crustaceans, worms, young leeches, snails, and occasional plant material. The male lives longer than the female and can attain an age of four years. D. S. Jordan and B. W. Evermann, two of America's earlier and premier ichthyologists, refer to it as the "darter of darters—the hardiest, wiriest, and wariest, and the one most expert in catching other creatures." It is perhaps one of the best darters to keep in an aquarium.

Carolina darter
Etheostoma collis Pl. 185
Swamp darter
Etheostoma fusiforme Pl. 186
Sawcheek darter
Etheostoma serrifer Pl. 187

Description. These three species are placed in the subgenus *Hololepis,* which contains a total of six species. The group is characterized by a usually incomplete lateral line that arches distinctly upward.

Carolina darter: 1.2 to 2.4 inches (31 to 60 mm). This small, nondescript darter is brown on the back and side and white to yellow below. The side is speckled with dark brown spots, there is a row of 8 to 14 dark brown dashes or blotches on the midside, and there are three black spots just before the base of the tail.

Swamp darter: 1.4 to 2.3 inches (36 to 59 mm). This is a small darter with a brown to brown-green compressed body, dark saddles on the back, 10 to 12 blackish blotches or squares along the side, a yellow, dark-flecked belly, a weak black bar below the eye, and three black spots just before the tail.

Sawcheek darter: 1.6 to 2.7 inches (40 to 68 mm). A serrated rear edge of the preopercle bone, located on the middle of the gill cover, is the basis for its common name. Similar to the swamp darter in appearance, it is brownish above and on the side, greenish below with black flecks, and has four vertically aligned black spots at the base of the tail, the middle two of which are surrounded with red. It often has vertically elongated dark brown blotches on the side.

Distribution and Abundance. The Carolina darter is restricted to the mid-Atlantic region and is found only in the piedmont from south central Virginia through North Carolina into north central South Carolina. The swamp darter in the region is found along the coast from central Delaware to the Savannah River in South Carolina; it has been introduced into the French Broad River in western North Carolina. It also occurs north to southern Maine, south and west along much of the Gulf coast, and in the lower Mississippi River basin. The sawcheek darter is found on the coastal plain from southeastern Virginia to eastern Georgia. All three species can be common, and the swamp darter is often abundant. Because of its restricted distribution, the Carolina darter is considered threatened in Virginia and of special concern in North Carolina.

Habitat. The Carolina darter occurs in sluggish to calm, clear to slightly turbid creeks and small rivers over a bottom of mud, sand, and rock, including bedrock. Unlike almost all other darters, the swamp darter prefers still, slow water, over mud or sand, usually in or near heavy aquatic vegetation. The sawcheek darter is usually found in small to medium-sized creeks with a more moderate current over a sand to gravel substrate and often also with submerged aquatic vegetation.

Natural History. Food of the Carolina darter is tiny crustaceans and larvae of aquatic insects, which are located by sight. It spawns in March and April, when the water temperature approaches 57°. The male courts the female by erecting his fins and then contacting her with his fins and head. He then mounts her, and he is sometimes joined by other males. Eggs are attached to gravel, sticks, leaves, and other substrates. Food of the swamp darter is small aquatic insects and amphipods. In the mid-Atlantic region the swamp darter spawns in March or April, and it apparently does not establish a territory. The male mounts the female from the rear and beats her with his pelvic fins studded with breeding tubercles. The pair then moves to underwater plants, where the eggs are laid singly on leaves. The maximum length of life is usually one year. The biology of the sawcheek darter is unknown, but it is probably similar to that of its two relatives.

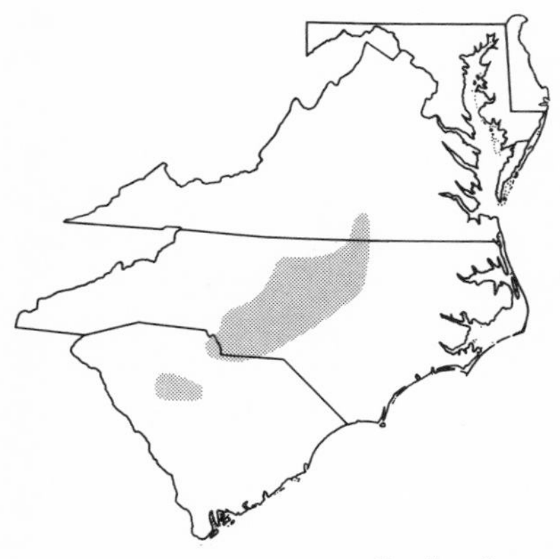

Carolina darter

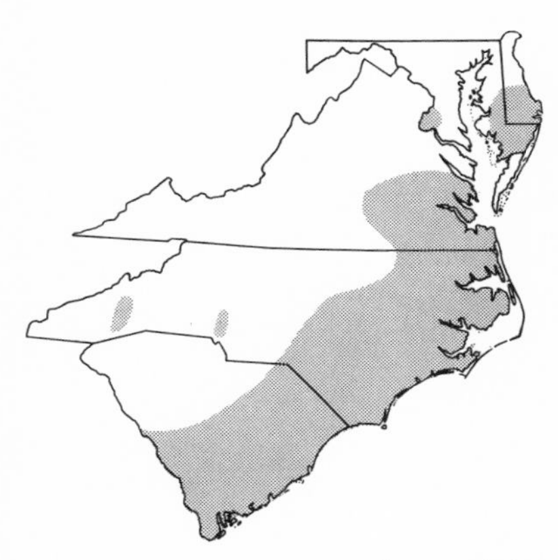

Swamp darter

Sawcheek darter

Yellow perch

Perca flavescens Pl. 188

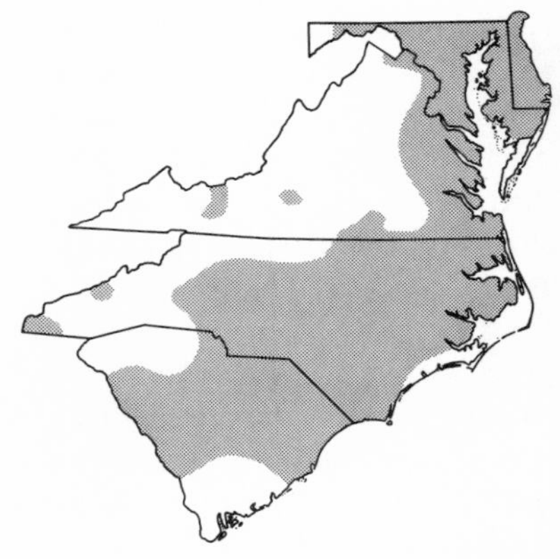

Yellow perch

Description. 6.0 to 15.7 inches (152 to 400 mm). The yellow perch is identified as a member of the perch family by the presence of two distinct dorsal fins, an anterior one with sharp spines and a posterior one that consists mostly of soft rays. The body is compressed and high and is usually tan-colored with a series of tall black bars along the side. The pectoral, pelvic, and anal fins are tinged with yellow or orange.

Distribution and Abundance. This fish is found across the northern United States and much of Canada, and its range extends south to include most of the mid-Atlantic region except the lower portion of South Carolina. It has been extensively introduced into lakes and reservoirs. It is often abundant, and this sometimes results in stunted individuals.

Habitat. The yellow perch is found in a variety of habitats, from clear, cool lakes to slow streams. It is tolerant of strongly acid waters and is regularly caught in blackwater lakes and streams of the coastal plain in the mid-Atlantic region. It also occurs in slightly brackish waters of the upper reaches of estuaries. It is generally intolerant of pollutants and heavy siltation.

Natural History. The yellow perch spawns at night in late winter or early spring in the mid-Atlantic region. Large numbers of eggs are produced in long, sticky strings or bands that adhere to aquatic vegetation or settle to the bottom. No care is given to the eggs, and upon hatching, the fry are totally independent. The male usually reaches maturity in its second year of life, and the female usually in the third year; individuals attain an age of 13 years. Food is primarily small fishes, insects, crayfishes, and other invertebrates. The yellow perch seldom weighs more than one pound in the mid-Atlantic region and consequently is not heavily fished for there, although elsewhere it reaches a weight of over four pounds. In the mid-Atlantic region it is usually taken incidental to sunfishes, on crickets or minnows, and with various spinners and spoons. The flesh is firm and makes excellent table fare.

Blackbanded darter

Percina nigrofasciata Pl. 189

Dusky darter

Percina sciera Pl. 190

Description. These two darters are included in the subgenus *Hadropterus,* which contains four species. This group is characterized by a vertical row of three dark spots on the tail fin base. The males of all members of the genus *Percina* have a row of modified scales on the midline of the belly.

Blackbanded darter: 1.8 to 4.3 inches (45 to 110 mm). It is characterized by an olive to black back with six to eight dark saddles, 12 to 15 high, dark to black bars on a light green or brown or white side, and dark wavy lines on the upper side. There is no, or only a faint, dark bar below the eye.

Dusky darter: 1.8 to 5.1 inches (46 to 130 mm). Of subdued colors, this fish has 8 to 12 rounded black blotches on the side, is olive to blackish above, has eight or nine dark brown saddles on the back, and has a wavy dark brown line on the upper side. There is usually no black bar under the eye.

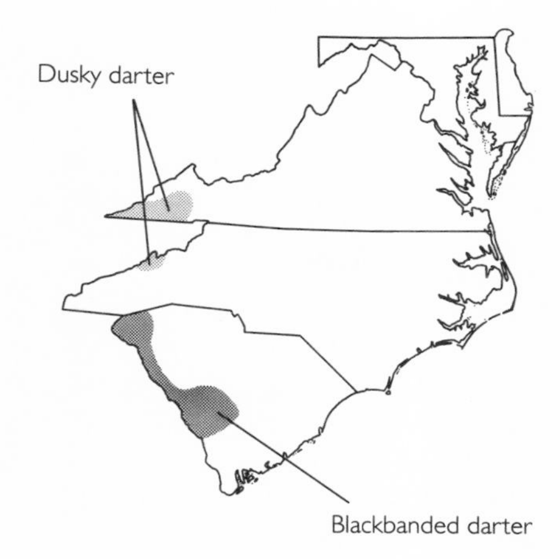

Distribution and Abundance. In the mid-Atlantic region the blackbanded darter is found in western South Carolina; it also occurs further south and west. Where present, it is often the most abundant darter. The dusky darter in the region is found only in extreme western North Carolina and far western Virginia, where it is usually rare; it is considered endangered in the former state. It also occurs from central Texas almost to the Great Lakes.

Habitat. Both of these darters occur in riffles and runs of small creeks to moderate-sized rivers, over a bottom of sand or gravel; the blackbanded darter also occurs in heavy vegetation and over mud. Both are inhabiters of pools, and both prefer a slow to moderate current.

Natural History. Food is sought visually during the daytime and consists primarily of immature aquatic insects. Food intake is greatest in April, May, July, and September. Both species spawn in May to early June over gravel in riffles and raceways. The males are probably territorial, and in both species the male is larger than the female. The female of the dusky darter produces some 80 to 200 eggs per season. Sexual maturity in both species is reached when one year old, most individuals do not survive a second spawning season, and only a rare individual lives for more than three years. The dusky darter probably overwinters in deeper downstream waters. This species is highly sensitive to turbidity, silt, damming, and other forms of pollution.

Sharpnose darter
Percina oxyrhynchus
Olive darter
Percina squamata Pl. 191

Description. These darters are in the subgenus *Swainia* and are separated from others by a submarginal orange band on the first dorsal fin and a black spot at the base of the caudal fin. Both have a long and pointed snout.

Sharpnose darter: 2.8 to 4.7 inches (70 to 120 mm). It is characterized by 10 to 12 dark brown rectangles along the side, an unscaled or partially scaled breast, and 13 to 15 small dark brown saddles.

Olive darter: 2.8 to 5.1 inches (70 to 130 mm). Nearly identical to the sharpnose darter, the olive darter has a fully scaled breast.

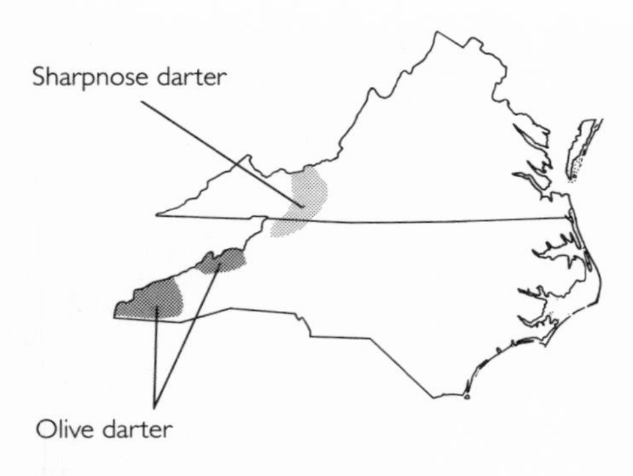

Distribution and Abundance. In the mid-Atlantic region the sharpnose darter is restricted to the New River drainage in northwestern North Carolina and southwestern Virginia; it is also found in other southern tributaries of the Ohio River from Pennsylvania to Kentucky. The olive darter in the mid-Atlantic region is restricted to tributaries of the upper Tennessee River in North Carolina; it is found also in tributaries of this river in Georgia and Tennessee, as well as in the Cumberland River drainage in Kentucky and Tennessee. Both species are usually localized but are sometimes abundant; both are listed as of special concern in North Carolina.

Habitat. Both species occur in fast riffles among boulders in small to medium-sized rivers, over a bottom of gravel and rubble.

Natural History. Little is known about these fishes. The sharpnose darter probably spawns in April and May, while the olive darter spawns in May and June.

Piedmont darter
Percina crassa Pl. 192
Longhead darter
Percina macrocephala
Stripeback darter
Percina notogramma
Shield darter
Percina peltata Pl. 193
Roanoke darter
Percina roanoka Pl. 194

Description. Six of nine species in the subgenus *Alvordius* occur in the mid-Atlantic region. The group is distinguished by the combination of separate branchiostegal membranes, a wide premaxillary frenum, and an absence of breeding tubercles and—except in the Roanoke darter—bright body colors. Identification, especially of the young and females, is usually difficult for the nonexpert.

Piedmont darter: 2.7 to 3.5 inches (68 to 90 mm). This is a brown fish with seven to nine oval black blotches on the side, a black bar on the chin, a large black spot on the breast, and a large black teardrop below the eye. The first dorsal fin has a black edge, a tan-orange band below, a yellow interior, and vertical black markings before the lower portion of each spine.

Longhead darter: 2.9 to 4.7 inches (73 to 120 mm). This species is characterized by a wide black lateral band with undulated edges, a long snout, a black bar below a black spot located on the base of the tail, and, below the eye, a sickle-shaped black bar that curves backward and extends to the underside of the head.

Stripeback darter: 2.0 to 4.3 inches (52 to 110 mm). This species is distinguished by a pale yellow stripe on the upper side of large individuals, a first dorsal fin that is dusky throughout, a black teardrop below the eye, and a black spot on the base of the caudal fin.

Shield darter: 2.7 to 3.5 inches (68 to 90 mm). This darter has a black bar on the chin, a black teardrop below the eye, a large black spot on the breast, a row of elongate black markings on the first dorsal fin, and a large black blotch just below the center of the base of the caudal fin. The **Appalachia darter**, *P. gymnocephala*, is similar, but it has black ovals on the first dorsal fin and lacks the black bar on the chin.

Roanoke darter: 1.6 to 3.1 inches (40 to 78 mm). The most colorful species of this group, it has 8 to 14 black bars on the side in the adult (which are oval or vertically narrow blotches in the young), a gray-blue back and gold-orange belly, and a wide orange band on the first dorsal fin in the adult male.

Distribution and Abundance. The Piedmont darter is found in only a small portion of southwestern Virginia, from northwestern to southeastern North Carolina, and in north central South Carolina. It is usually common inland and uncommon in the coastal plain. In the mid-Atlantic region, the longhead darter now occurs only in western Virginia, in which state it is considered to be threatened, as it has probably been extirpated from its mountain range in North Carolina. It also occurs in adjacent states to the west and north, where it is rare and usually local, and it has disappeared from many areas. The stripeback darter is restricted to most of central and northern Virginia and west central Maryland and, outside of the mid-Atlantic region, to extreme eastern West Virginia. It is common, except in Maryland where it is listed as endangered. The shield darter occurs in central North Carolina, most of Virginia and Maryland, and northern Delaware within the mid-Atlantic region, and then north to central New York. It is one of the region's most common darters. The Appalachia darter is restricted to the New River drainage in western North Carolina and Virginia and in southeastern West Virginia. The Roanoke darter is found in much of central North Carolina and central to southeastern Virginia, and it was probably recently introduced in parts of the latter state. Its range extends into adjacent southeastern West Virginia, which it apparently entered recently from Virginia. It is locally common to abundant, especially in the mountains and piedmont, and less numerous on the coastal plain.

Habitat. All of these species live in clean water with a moderate current and usually a stony bottom free of vegetation and mud. The longhead and stripeback darters live in pool areas, and often in silty areas, of larger creeks and small to medium-sized rivers. They are usually found among sunken brush (twigs). The other three species prefer faster-flowing rock to gravel riffles of creeks or small and medium-sized rivers.

Natural History. Little is known of the natural history of this group. All are probably active only during the day and search for food, shelter, and mates by sight. All feed primarily on immature aquatic insects, and the longhead darter also feeds on crayfishes. Most of these species probably spawn in March and April, although the shield and Roanoke darters spawn in April and May. Ripe females of the stripeback darter have been found in waters of 46° to 61°, and eggs of the shield darter are said to ripen at a water temperature of 50°. The shield darter spawns primarily in areas of shallow water over gravel in calmer water; the male, which lacks tubercles, establishes a stationary territory, as well as a moving territory that is centered around a female. The male of all species probably attempts to drive off rivals and to attract or to remain near a female. When mating, the male mounts the female, both remain in a horizontal position, and both quiver as they expel

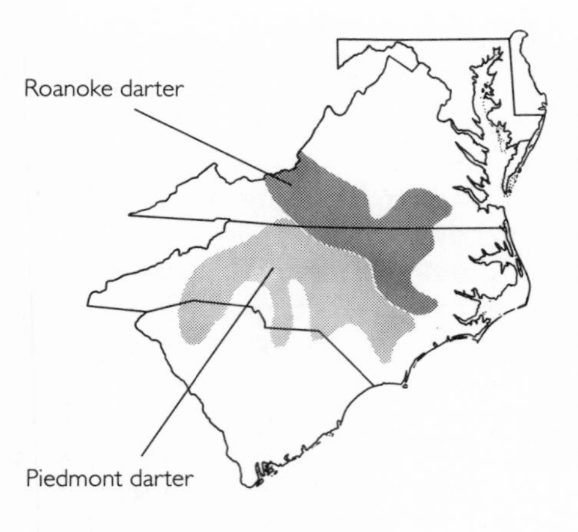

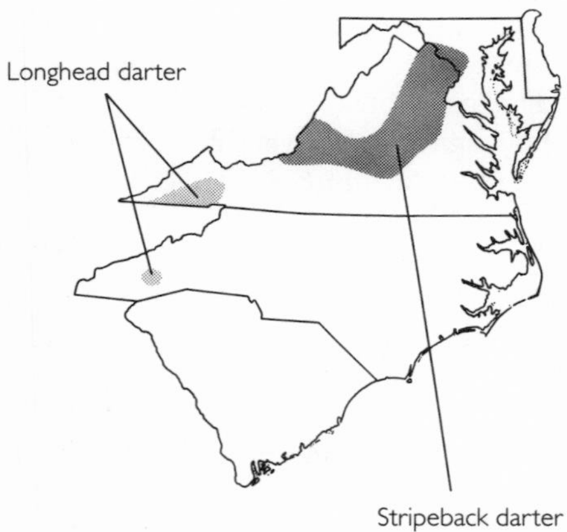

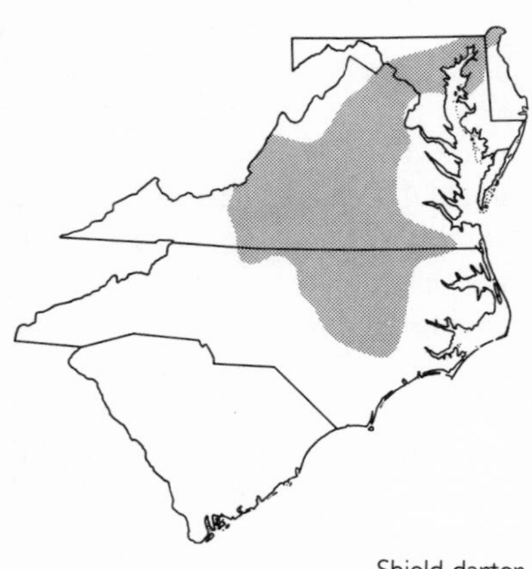

gametes. The fertilized eggs are buried in gravel and are not guarded. The young grow rapidly and probably live in calmer waters than do the adults. The shield darter schools loosely in beds of aquatic vegetation in summer and autumn. The shield darter matures at an age of one year, while the longhead darter matures at two years. The Roanoke darter attains a known maximum age of three years, while the longhead darter attains four years.

Gilt darter

Percina evides Pl. 195

Description. 2.3 to 3.8 inches (58 to 96 mm). This colorful fish is characterized by six to eight wide dusky green or black saddles that join similar-colored blotches on the side. The underside, and sometimes most of the body, is of a bright yellow to orange color. There is an orange band below the edge of the first dorsal fin and often another on the fin base, there is a large black teardrop mark below the eye, and there is a yellow-orange iridescent line just below the teardrop. Colors in the breeding male are intensified. This is one of two species in the subgenus *Ericosma*, which is characterized by tubercles on much of the belly of the breeding male and by the bright colors of the adults.

Distribution and Abundance. The gilt darter in the mid-Atlantic region occurs in the Tennessee River drainage of North Carolina and Virginia. It is widely distributed in disjunct populations in much of the east central United States. Often still locally abundant, it has been extirpated from much of its former range, including all of several states, due to habitat degradation.

Gilt darter

Habitat. This beauty occurs in clear, medium-sized to large creeks and small and medium-sized rivers, in moderate to fast deep riffles and pools over sand, gravel, rubble, and boulders.

Natural History. The gilt darter feeds on immature aquatic insects and snails, which it finds by carefully searching around, under, and between rocks in its habitat. It winters in deeper pools, and returns in spring to shallow water over riffles. In the mid-Atlantic region it spawns in May at a water temperature of between 63° and 68°. The male is larger than the female. When ready to spawn, the female swims over the substrate and searches for a suitable spawning site, closely followed by the male. Once a site is located, she rests on the bottom and is then mounted by the male. As vibrations of the pair become intense, sperm and eggs are released and are pushed into the sand and gravel below. The species first spawns at an age of one year and can live for a maximum of four years. Older males die shortly after they have spawned.

Tangerine darter

Percina aurantiaca Pl. 196

Description. 4.3 to 7.1 inches (109 to 180 mm). The only member of the subgenus *Hypohomus*, the tangerine darter has a broad black stripe of 8 to 12 fused blotches on the side, a row of small brown spots on the upper side, a generally gray-brown or olive dorsum, and a thin black stripe along the back, which breaks into spots at the rear. The lower side and the venter of the young are white; they are yellow in the mature female and tangerine red in the mature male. This species lacks the black teardrop bar under the eye typical of most members of the genus *Percina*, and the male lacks the large belly scales present in the other species of the genus. An adult male rivals any fish species in bright coloration.

Distribution and Abundance. This species is found only in the Tennessee River drainage, and in the mid-Atlantic region occurs only in western Virginia and North Carolina. It also occurs in adjacent Georgia and Tennessee. It is rare to common.

Habitat. The tangerine darter inhabits large creeks and small rivers with clear water, in deep bedrock and boulder riffles with an extremely rapid current. The young and the female are sometimes found in calmer water.

Natural History. The food of the tangerine darter is primarily immature

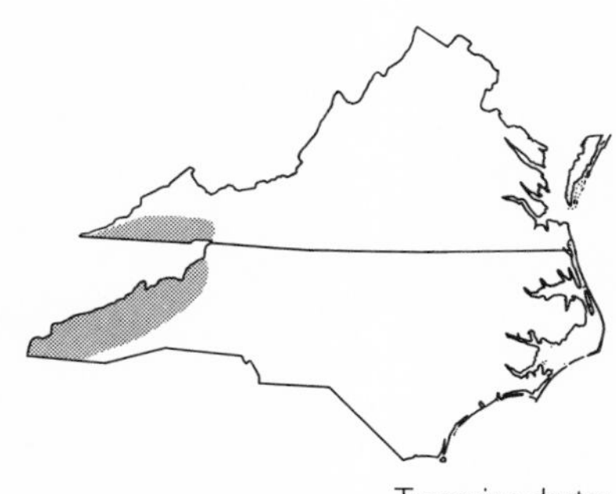

Tangerine darter

aquatic insects, which it obtains by nipping at them. This fish is most active in the early afternoon. It spawns from May to July, and the male then becomes territorial. The number of ripe eggs found in females ranges from 5 to 750. The female is mounted by the male, and both quiver and simultaneously release eggs and sperm. A few eggs are released with each mating, and the process is repeated. The eggs are not cared for, unlike in many other darter species. Some males become mature when one year old, and the female when two years old. The species attains an age of at least four years.

Channel darter

Percina copelandi Pl. 197

Description. 1.6 to 2.9 inches (42 to 73 mm). The only member of the subgenus *Cottogaster*, the channel darter has a slender body that is brown-gray on the upper half and silver-white below. It can also be identified by nine or ten small lateral blotches on the side, a blunt snout, a black spot on the midline of the base of the caudal fin, and numerous black X- and W-shaped marks on the back and upper side. The first dorsal fin of the male is black on the margin and on the base.

Distribution and Abundance. This species has a large and disjunct distribution. It occurs sporadically from the St. Lawrence River drainage in Canada to the southwestern Mississippi River basin, and in the mid-Atlantic region it occurs only in the Clinch and Powell rivers in southwestern Virginia. It is rare in Virginia, although it is sometimes common elsewhere.

Habitat. This darter inhabits only rivers in the mid-Atlantic region, primarily those that are warm and with a low to moderate gradient. In the northern parts of its range it also occurs in lake shallows.

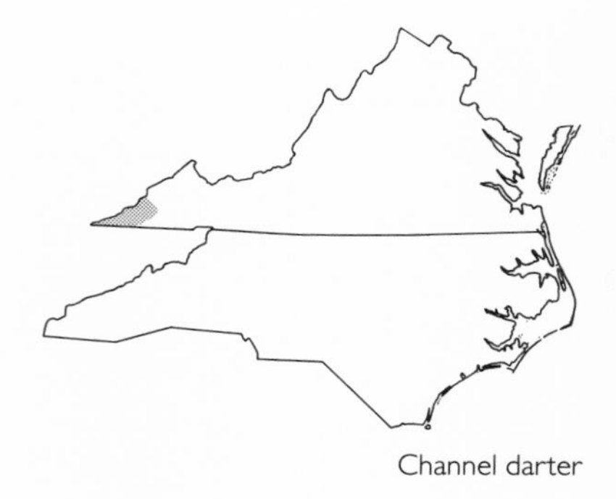

Channel darter

Natural History. The channel darter spawns in April and May, when the water warms to 69° to 72°. The male establishes a territory that is centered around at least one large rock. The ripe female enters his territory, is directed toward the center, burrows into gravel behind the rock, and is then mounted by the male. Four to ten eggs are then buried in gravel per spawning session. There is no parental care of the eggs. Maturity is reached at one year of age, and the male grows larger than the female. Food is primarily larvae of mayflies and midges, as well as microcrustaceans.

Blotchside logperch

Percina burtoni Pl. 198

Logperch

Percina caprodes Pl. 199

Roanoke logperch

Percina rex Pl. 200

Description. Three of the six species of logperches, all in the subgenus *Percina*, occur in the mid-Atlantic region. All have a distinct head shape and color pattern: there is a wide, flat area between the eyes, a row of round blotches or vertical bars on the side, and a black spot on the base of the caudal fin; the snout is pointed or bulbous (sometimes piglike), extending well beyond the upper jaw, and the snout tip is connected to the upper lip. Only the male has breeding tubercles.

Blotchside logperch: 4.2 to 6.3 inches (108 to 160 mm). Scales are absent or partly cover the nape, and the body is yellow-brown above and yellow to white below. It has eight to ten dark green to black round or oval blotches on the midside, and the first dorsal fin has an orange edge. The breeding male is the most colorful of the logperches.

Logperch: 4.7 to 7.1 inches (120 to 180 mm). The body is yellow-brown above, and the pattern on the side is of 16 to 22 narrow and alternating long and short bars that extend over the back and down the opposite side.

Roanoke logperch: 3.8 to 5.9 inches (96 to 150 mm). The body is dark olive to yellow-brown above, with wavy dark blotches and with ten to twelve short black bars on the side, which do not join those on the opposite side; there is a bold dark bar beneath the eye. A reddish-orange band is present near the edge of the first dorsal fin.

Distribution and Abundance. The blotchside logperch is known from the higher elevations of the Tennessee and Cumberland rivers of Virginia, North Carolina, Kentucky, and Tennessee, where it is rare. It may recently have been extirpated from the Cumberland River. It is considered endangered in North Carolina. The logperch is widespread from Canada to Gulf of Mexico drainages and in Atlantic slope drainages from the Hudson River in New York to the Susquehanna River in Maryland. It is common throughout most of its range, but it is peripheral and generally uncommon in the mid-Atlantic region. It is listed as threatened in North Carolina. The Roanoke logperch is scarce in the upper Roanoke, upper Dan, and upper Chowan rivers in Virginia, where it is protected as an endangered species as designated by the federal government.

Habitat. All three logperches occur in rubble and boulder runs and riffles of clear and small to medium-sized rivers. The logperch also occurs widely in other habitats, such as vegetated lakes and reservoirs.

Natural History. The logperch has been widely studied, and less is known about

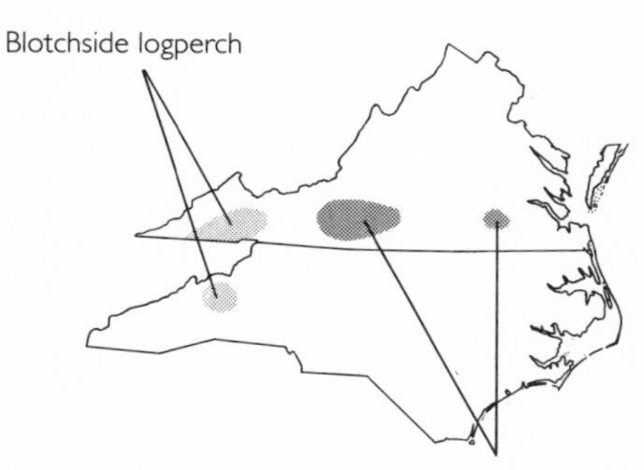

the other two species. In the mid-Atlantic region the logperch begins to spawn in spring when the male moves into shallow water, where it may gather in schools. Females ready to spawn swim through the males and are followed by one or two of them. When the female stops over sand, she is mounted by a male; both vibrate and burrow slightly into the substrate, and the female releases 10 to 20 adhesive eggs. The female mates with more than one male. The eggs are not guarded, they hatch in five to seven days, and the hatchling is free swimming at a length of a quarter inch. Maturity is reached during the second year, and the logperch lives to a maximum of about four years; the Roanoke logperch attains six years. There is no apparent difference in size between the sexes at any age.

The very young logperch feeds near the surface on small crustaceans, and the adult of all three species eats the larvae of midges, mayflies, and caddisflies. All three species actively search for food under stones and waterlogged pieces of wood, for which a long snout with a pad of thick protective tissue at the front is an adaptation. However, the following observations have been made on the blotchside perch. Stones or wood are upended or flipped over by the fish, using its body as a lever. Organisms found attached to the underside, or found underneath the object, are quickly swallowed. The stones range in maximum length from about one-half inch to 4½ inches and weigh up to eight ounces. A fish seems to know which objects are of a size suitable for it to move, and only occasionally does a fish attempt to lift one that is too heavy for it. Logperches are eaten by walleye, largemouth bass, rock bass, birds, and river otter, and they are sometimes caught by fishermen who fish with small worms as bait.

Sauger

Stizostedion canadense Pl. 201

Description. 10.0 to 30.0 inches (255 to 760 mm). While similar to the yellow perch, the sauger is longer and more slender, and has an elongate and pointed snout, a larger mouth, and large canine teeth. The dorsum is olivaceous, and there is a series of three or four elongate dark blotches on the upper side, which are fewer in number, wider, and not as high as in the yellow perch. There are two or three rows of half-moon-shaped black spots on the membranes between the spines of the first dorsal fin, three or four rows of weak black spots on the second dorsal fin, and four columns of black dashes on the caudal fin. The typical adult sauger usually weighs about two pounds, although some individuals reach ten.

Distribution and Abundance. The sauger is native to north central North America, and it apparently did not occur east of the Appalachian Mountains until it was introduced there. It has been introduced into reservoirs in the mid-Atlantic region as far south as the Savannah River drainage in South Carolina.

Sauger

Habitat. The sauger is typical of large reservoirs and rivers. It spends most of the year in open and deep water.

Natural History. This species moves from the open water of lakes, reservoirs, and rivers into the flowing water of tributary streams in early spring to reproduce. It spawns at night in shallow water. A female can lay up to 210,000 eggs. No care is given to the eggs, which hatch in about two weeks. Maturity is reached in the male at two years and in the female at three. Few live beyond five years, and the approximate maximum age attained is eight years. The fry and juvenile eat a variety of aquatic invertebrates, while the adult is primarily piscivorous. The sauger is a popular game fish, best caught at night in deep water during the warmer months. Live bait (minnows) or spinners and other artificial lures that imitate fishes and that are worked slowly usually produce good results; nightcrawlers are sometimes also effective. It is an excellent eating fish.

Walleye

Stizostedion vitreum Pl. 202

Description. 11.8 to 35.8 inches (300 to 910 mm). An opaque and whitish eye when the fish is dead gives this large perch its name. The walleye is shaped like a sauger. An olive dorsum is broken by some five dark saddles that extend down onto the upper side, and the olive dorsum color shades to white on the belly. The lower lobe of the caudal fin is white at the tip. This feature and a large dark spot at the posterior base of the otherwise uniformly colored first dorsal fin help to separate it from its close relative, the sauger.

Distribution and Abundance. The walleye is widespread across most of Canada and the northern United States. Its status as native in the the mid-Atlantic region is uncertain. It has been widely stocked and now occurs in lakes and reservoirs in the piedmont and mountains of the region as far south as South Carolina. Its abundance varies greatly from lake to lake.

Habitat. This fish is intolerant of pollution and heavy siltation. It usually occurs in large, clear, cool lakes and rivers, in relatively deep water. It may occur in shallow water during the cooler parts of the year, but it moves to deeper areas as temperatures rise in summer.

Natural History. The walleye migrates from open lake waters into tributary streams in late winter or early spring when the water temperature rises to 42° to 50°. Here it spawns over sand or gravel bars at night. The female broadcasts her eggs, which are then fertilized by sperm released by attendant males. The eggs then sink and adhere to rocks and sand. A female can lay up to 495,000 eggs. No care is given to the developing embryo, which hatches in from five days to two weeks. The male matures at five

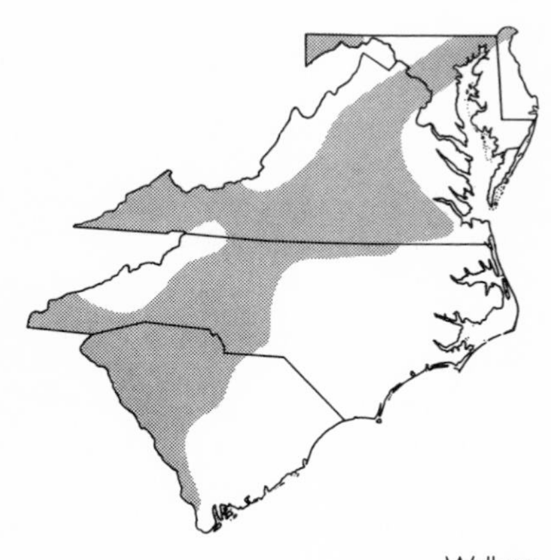

Walleye

years of age and the female at six, and the maximum age attained is 29 years. The fry eats a variety of aquatic invertebrates, while the adult is primarily a fish eater. The walleye is most active in the evening, when it moves onto bars and shoals to feed. Fishing techniques are as described for the sauger above. The flesh is firm and white, and it is considered by many to be the best eating of all freshwater fishes.

Drums *Family Sciaenidae*

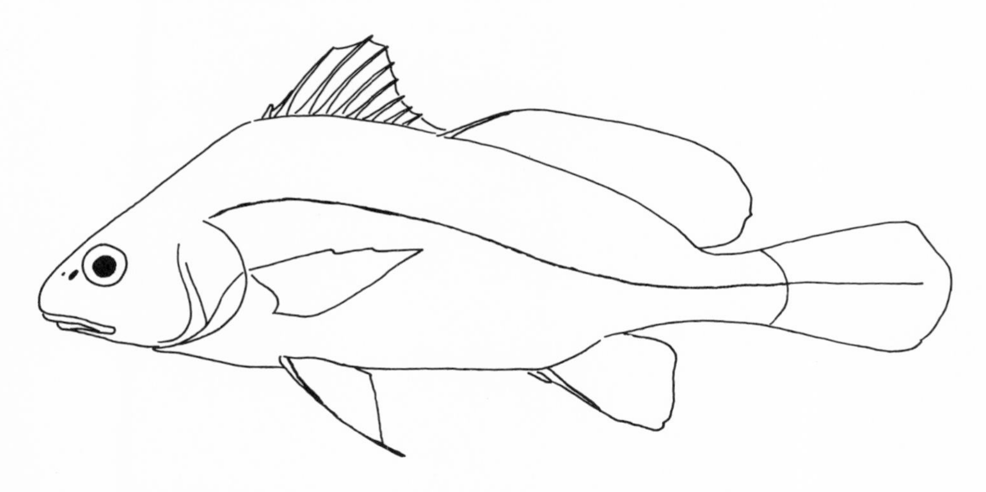

This family of about 210 species contains some 50 genera found in nearshore marine habitats off all the continents, as well as in brackish water and fresh water. In North America, only a few ascend into fresh water and only one is restricted to it. All have a long dorsal fin, usually with a deep notch that separates the anterior spinous portion from the posterior soft portion, and a complete lateral line. Some reach a length of several feet. Most forms in the United States have a rounded snout, large eye, inferior mouth, and flattened underside. Many can make a sound with their air bladder, hence the common names "drums" and "croakers." Most are benthic and feed on mollusks, worms, and crustaceans, although some are midwater predators on fishes. Many are important food and game fishes.

Freshwater drum

Aplodinotus grunniens Pl. 203

Description. 12 to 39 inches (305 to 1,000 mm). The body is strongly compressed and humpbacked, the venter is flattened, the body is gray-white or silvery, the snout is rounded, and the mouth is horizontal. The anterior spinous dorsal fin is joined to the much longer soft dorsal fin. The edge of the tail is rounded or slightly pointed, and the scales are ctenoid. There are two spines on the anal fin, of which the second is much larger than the first.

Distribution and Abundance. This drum occurs in the mid-Atlantic region only in far western Virginia and extreme western North Carolina. Overall, it occurs from the Great Lakes and adjacent Canada south through most of the Mississippi River basin to western Texas and extreme western Florida. While still abundant to common in much of its range, it has been reduced in numbers in parts of the range and eliminated in others. Because of its limited distribution in North Carolina, it is there considered to be threatened.

Habitat. This drum occurs primarily in large rivers, but it is also found in large lakes and impoundments and sometimes in small rivers. It prefers open areas of warm and sluggish water, and

Freshwater drum

turbid rather than clear waters. It is a bottom dweller and favors a bottom of mixed sand and silt.

Natural History. Although some authors report this species to feed extensively on mollusks, which would seem appropriate because of the large and powerful molarlike teeth in its pharynx, more recent studies suggest that the adult feeds mainly on insects, crayfishes, and fishes; the young prefers benthic insects and minute crustaceans. This drum feeds at all hours, and moves into shallow waters at twilight. It feeds primarily by touch and taste and often moves rocks with its snout to disclose food organisms. It shows an early tendency to school, which continues throughout life; it sometimes forms large concentrations.

The spawning season lasts from May to July and begins at a water temperature of 66°. The male then produces drumming sounds, which are probably related primarily to reproduction. It spawns in open water, usually far from shore. Several hundreds of thousands of eggs are laid per female, and they and the larvae float on the water surface rather than sink. This floating is something that is unique to North American freshwater fishes, but it is typical of many marine species. The male can mature at an age of two years and the female at four, and the species can reach an age of many years. While there are numerous recent records of drums weighing fifty pounds or more being caught, in pre-Columbian times this fish may have attained a weight of up to 200 pounds. In some areas it is sought for the sport provided by its fighting abilities and its large size, as well as for its edible flesh; it is sometimes served in restaurants.

Glossary

Adipose eyelid: Fatty translucent tissue that totally or partially covers the eye in some fishes.

Adipose fin: Fleshy, small fin that lacks rays and spines and is located on the dorsum, just before the tail.

Air bladder: Swim bladder or gas-filled bladder inside the body, just below the backbone; it regulates hydrostatic pressure and in some fishes is an auxiliary respiratory structure.

Ammocoetes: Larval stage of a lamprey.

Amphipod: Member of an order of small crustaceans with a laterally compressed body (e.g., scud).

Anadromous: Moving from the ocean into fresh water to spawn.

Anal fin: Median fin located on the ventral side, just behind the anus and anterior to the caudal (tail) fin.

Anal fin spine: Spine located at the front of the anal fin; it may be short and embedded in the flesh.

Anterior: Of or pertaining to the front.

Anus: Posterior opening of the digestive tract.

Arthropod: Member of a phylum of invertebrates that includes insects, crustaceans, and spiders.

Axillary process: Narrow flap of flesh located just above the base of the pectoral or pelvic fins.

Band: Wide color mark(s) on a fin.

Bar: A vertical color mark on the body or fin.

Barbel: Fleshy projection around the mouth, nostrils, or chin in some fishes; it may be large and obvious or small and inconspicuous.

Biota: All the animal and plant life of a particular region.

Body depth: Maximum vertical straight-line dimension of the body, excluding the fins.

Branchiostegal (gill) membrane: Membrane attached to the lower edge of the gill cover, which helps to enclose the gill chamber; it is supported by thin and elongate bony elements called branchiostegal rays.

Breeding tubercle: *See* Tubercle.

Bryozoans: Members of a phylum of small aquatic animals that form mosslike colonies called moss animals.

Canine teeth: Long, sharp teeth (as in pickerel) for piercing.

Catadromous: Moving from fresh water to the ocean to spawn.

Caudal fin: Tail fin.

Caudal peduncle: Narrow posterior part of a fish that connects the tail to the body.

Cladoceran: Member of a suborder of small crustaceans called water fleas, including *Daphnia.*

Copepod: Member of a subclass of small crustaceans similar to amphipods.

Crustacean: Member of a class of arthropods, including shrimps and crayfishes.

Ctenoid scale: Thin, light, flexible scale with numerous small "teeth" that point backward.

Cycloid scale: Thin, light, flexible scale that lacks small, rearward-pointing "teeth."

Decurved: Curved downward.

Demersal: Sinking to the bottom.

Detritus: Decomposed plant and animal material.

Diatom: Member of a group of microscopic single-celled alga with a hard shell of silica.

Distal: Of or pertaining to the point most remote from the place of attachment.

Diurnal: Of or pertaining to the daytime.

Dorsal: Of or pertaining to the back or upper part of the body.

Dorsal fin(s): The medial fin(s) of the midline of the back; in many fishes there are two, often consisting of a spiny-rayed anterior dorsal fin and a posterior soft-rayed dorsal fin, which may or may not be connected.

Dorsal fin spine: Spine located on the dorsal fin.

Dorsum: Back or upper part of the body.

Elver: Juvenile stage of the American eel.

Enamaloid teeth: Teeth that are covered with enamel.

Endemic: Restricted in occurrence to a particular and usually relatively small area.

Eutrophication: Accumulation of nutrients in lakes and other bodies of water that causes rapid growth of vegetation; it usually results from the activities of humans.

Exotic: Not native to a given area or region.

Exterminated: Condition in which all individuals of a species are dead because of the activities of humans.

Extinct: Condition in which all individuals of a species are dead, either as a result of the activities of humans or for other reasons.

Extirpation: Complete removal from a portion of the range of a species, owing to the activities of humans.

Fall line: Boundary area between the piedmont and coastal plain.

Frenum: Fleshy bridge or connection between the snout and upper lip.

Fry: Young, newly hatched fish.

Fusiform: Streamlined, cigar-shaped, tapering at both ends (refers to body shape).

Gamete: Mature reproductive cell; sperm and egg.

Ganoid scale: Thick, inflexible scale covered with hard enamel, such as on gars.

Genital papilla: Small, fleshy projection at the genital pore immediately behind the anus; it may be bifurcate (forked) or single and broad.

Genus: Taxonomic category that includes a single species or a closely related group of species (plural = genera).

Gill: Gas-exchange organ of fishes found on either side of the body under the gill cover or, in lampreys, in a pouch deep within the body wall.

Gill arch: Bony support and major component of the gills, to which gill rakers and gill filaments are attached.

Gill cover: Bony, flaplike cover posterior to the cheek, which protects the gills; also known as the operculum or opercle.

Gill filament: Soft, fleshy, threadlike structure attached to the posterior side of each gill arch.

Gill opening: External opening to the cavity that contains the gills.

Gill raker: Coarse and toothlike or long, fine, and hairlike structure found on the anterior edge of a gill arch.

Gonopodium: Modified and enlarged rays of the anal fin of a male livebearer (poeciliid) that serve as an intromittent (reproductive) organ.

Gular plate: Large, hard plate found on the chin in bowfins.

Herbivore: Plant eater.

Heterocercal: Of or pertaining to a caudal fin in which the vertebral column extends into the upper lobe and in which the upper lobe is usually longer than the lower lobe.

Ichthyologist: Scientist who studies fishes.

Isopod: Member of an order of crustaceans with a flattened body, such as pill bug.

Keel: Sharp ridge along the ventral midline, as in some herrings.

Keratinoid teeth: Teeth covered with keratin, a horny substance; found in lampreys.

Lateral line: Dotted line of pores on the body containing sense organs that can detect a change in water pressure.

Lateral scale: Scale on the midside, usually along the lateral line and usually considered as a row of scales that extends from the rear edge of the gill cover to the base of the caudal fin.

Maxillary: Bone of the upper jaw, located directly behind the premaxillary and sometimes bearing teeth.

Median fin: Any unpaired fin located on the midline of the body: the dorsal, anal, and caudal fins.

Milt: Sperm cells and seminal fluid produced by a male fish.

Mollusk: Member of a phylum of invertebrate animals with a hard shell, such as clams, mussels, and snails.

Morph: The form of an animal or plant species as characterized by its shape, size, or color.

Nape: Anterior portion of the back located immediately behind the head.

Natal: Of or pertaining to birth.

Nocturnal: Of or pertaining to the night.

Nostril: Nasal opening; fishes usually have two on each side.

Nuptial tubercle: *See* Tubercle.

Omnivore: An eater of food of both plant and animal origin.

Opercle: Principal bone of the gill cover; its posterior edge may be spiny, serrate, or smooth (entire). The term is sometimes used as a synonym for gill cover.

Opercular lobe: More or less conspicuous extension on the posterior edge of the operculum; it can be soft or stiff.

Operculum: *See* Gill cover.

Oral disc: Fleshy circular structure surrounding the mouth of a lamprey.

Origin: Point on the body where a fin begins.

Ova: Eggs, produced by the female.

Papilla: Small, fleshy protuberance, usually found on the lips.

Pectoral fin: Paired fin located high or low on the side and on the anterior portion of the body.

Pelvic fin: Paired fin located on the ventral side of the body, often called the ventral fin; it may be placed posterior to the pectoral fin (abdominal), below the pectoral fin (thoracic), or before the pectoral fin (jugular).

Pelvic girdle: Bones that support the pelvic fins.

Peritoneum: Lining of the abdominal cavity.

Pharyngeal teeth: Teeth on the bony arches in the throat area of some fishes.

Pharynx: Cavity at the back of the mouth that leads to the stomach.

Phytoplankton: Microscopic plant that floats or drifts in the water.

Piscivore: Organism that feeds on fishes.

Planktivore: Organism that feeds on plankton.

Plankton: Small animal or plant that floats or drifts in the water.

Plicae: Tight folds of skin on the lips.

Posterior: Of or pertaining to the rear.

Premaxillary: The most anterior paired bones in the upper jaw.

Preopercle: I-shaped bone located on the front portion of the gill cover.

Ray: Supporting rod of a fin that is composed of many small parts; it is usually branched and segmented but is never sharp.

Redd: Depression dug in a substrate in preparation for spawning; it is often referred to as a nest.

Riffle: Fast-flowing and shallow part of a stream located over rocks or gravel where the water surface is rough or rippled.

Roe: Fish eggs.

Rostrum: More or less elongated upper jaw.

Saddle: Color mark, often rectangular in shape, located on the back.

Scute: Scale that is enlarged, hardened, and sometimes pointed.

Septum: Thin dividing membrane.

Serrate: With a sawtooth edge.

Slough: Swampy or marshy aquatic habitat.

Snout: Part of a fish from the anterior tip of the upper jaw to the anterior part of eye.

Spine: Hard, bony, and unbranched fin support, often pointed.

Spiral valve: Spiral-shaped organ located inside the walls of the intestine; found in certain "primitive" fishes.

Stripe: Horizontal band of color on the body.

Subopercle: Bone of the gill cover located ventral to the opercle.

Subterminal mouth: Mouth located at a point below the anteriormost portion of the head.

Superfetation: Conception that occurs before the birth of the young that have resulted from a previous conception.

Sympatric: Two or more species or populations that exist in the same region.

Taxon: A taxonomic category (e.g., phylum, order, family, genus, species).

Total length: Length from the tip of the snout to the most distal portion of the caudal fin.

Tubercle: Small, hard protuberance on the skin, usually present only on a breeding male.

Venter: The belly or lower part of the body.

Ventral: Of or pertaining to the belly or lower part of the body.

Vermiculation: Color pattern of short and wavy (wormlike) lines.

Villi: Thin, elongate projections that grow out of a membrane.

Villiform: Small teeth set in bands; they are short, close together, and sandpapery to the touch.

Vomer: Bone located in the front of the roof of the mouth.

Selected References

Becker, G. C. 1983. *Fishes of Wisconsin.* Madison: University of Wisconsin Press. 1052 pp.

Burkhead, N. M., and R. E. Jenkins. 1991. Fishes. In *Virginia's Endangered Species*, K. Terwilliger, coordinator. Blacksburg, Va.: McDonald and Woodward Publishing Co. Pp. 321–409.

Jenkins, R. E., and N. M. Burkhead. 1994. *Freshwater Fishes of Virginia.* Bethesda, Md.: American Fisheries Society.

Kuehne, R. A., and R. W. Barbour. 1983. *The American Darters.* Lexington: University of Kentucky Press. 177 pp.

Lee, D. S., C. R. Gilbert, C. H. Hocutt, R. E. Jenkins, D. E. McAllister, and J. R. Stauffer, Jr. 1980. *Atlas of North American Freshwater Fishes.* Raleigh: North Carolina State Museum of Natural History. 854 pp.

Lee, D. S., S. P. Platania, C. R. Gilbert, R. Franz, and A. Norden. 1981. A Revised List of Freshwater Fishes of Maryland and Delaware. *Proceedings of the Southeastern Fishes Council* 3 (no. 3):1–10.

Loyacano, H. A., Jr. 1975. *A List of Freshwater Fishes of South Carolina.* Clemson: South Carolina Agricultural Experiment Station, Bulletin 580. 8 pp.

Manooch, C. S., III. 1984. *Fisherman's Guide—Fishes of the Southeastern United States.* Raleigh: North Carolina State Museum of Natural History. 362 pp.

Menhinick, E. F. 1991. *The Freshwater Fishes of North Carolina.* Raleigh: North Carolina Wildlife Resources Commission. 227 pp.

Nelson, J. R. 1984. *Fishes of the World.* 2nd ed. New York: John Wiley and Sons. 523 pp.

Page, L. M. 1983. *Handbook of Darters.* Neptune City, N.J.: T.F.H. Publications, Inc.. 271 pp.

Page, L. M., and B. M. Burr. 1991. *A Field Guide to Freshwater Fishes.* Boston: Houghton Mifflin Co. 432 pp.

Pflieger, W. L. 1975. *The Fishes of Missouri.* Jefferson City: Missouri Department of Conservation. 343 pp.

Quinn, J. R. 1990. *Our Native Fishes.* Woodstock, Vt.: The Countryman Press. 242 pp.

Raasch, M. S., and V. L. Altemus, Sr. 1991. *Delaware's Freshwater and Brackish Water Fishes.* Dover: Delaware State College. 166 pp.

Robins, C. R., R. M. Bailey, C. E. Bond, J. R. Brooker, E. A. Lachner, R. N. Lea, and W. B. Scott. 1991. *Common and Scientific Names of Fishes from the United States and Canada.* American Fisheries Society Special Publication 20. 183 pp.

Smith, C. L. 1985. *The Inland Fishes of New York State.* Albany: New York State Department of Environmental Conservation. 522 pp.

Smith, P. W. 1979. *The Fishes of Illinois.* Urbana: University of Illinois Press. 314 pp.

Trautman, M. B. 1981. *The Fishes of Ohio.* Columbus: Ohio State University Press. 782 pp.

Photo Credits

The photographs in this book were taken by Dr. James F. Parnell, except for those provided by the following photographers and agencies: Dr. Rudolf G. Arndt (borrow pit, Nassawango Creek, Red Clay Creek, pumpkinseed nests, lined topminnow), Mr. Richard G. Biggins, U.S. Fish and Wildlife Service (Cape Fear shiner), Mr. Richard T. Bryant (mooneye, spotfin chub, silverjaw minnow, streamline chub, rosyface chub, rosyface shiner, suckermouth minnow, Tennessee dace, fathead minnow, highfin carpsucker, muskellunge, brook trout, white bass, white crappie, Tippecanoe darter, sauger, walleye, freshwater drum), Mr. Noel M. Burkhead (Atlantic sturgeon, pearl dace, candy darter, channel darter, Roanoke logperch), Mr. Steven R. Layman (turquoise darter, Savannah darter), Dr. David G. Lindquist (Lake Waccamaw, kick seining, madtom eggs, Waccamaw killifish), Mr. Fred C. Rohde (bowfin, bluntnose minnow, sandhills chub, broadtail madtom, golden topminnow, least killifish, longear sunfish, Everglades pygmy sunfish, turquoise darter, Savannah darter, swamp darter), Dr. Stephen T. Ross (pugnose minnow), Dr. William N. Roston (bluehead chub, paddlefish, bowfin, rosyside dace, common carp, thicklip chub, warpaint shiner, greenhead shiner, yellowfin shiner, black crappie, tangerine darter), and Mr. J. R. Shute (thicklip chub, Watauga River, dusky shiner, northern studfish, bluefin killifish, blackbanded sunfish, bluespotted sunfish, dollar sunfish, Waccamaw darter).

Index